FUZZY LOGIC:
THEORY AND APPLICATIONS

Edited by
Aldon Cris B. Galido

Delve Publishing

Fuzzy logic: Theory and Applications

Edited by Aldon Cris B. Galido

Edition 2018

Delve Publishing

Published by
Delve Publishing LLC
708 3rd Avenue, 6th Floor
New York
NY 10801
United States of America
www.delvepublishing.com

ISBN: 978-1-68095-673-3 (Hardback)

Printed in United States of America

Library of Congress Control Number: 2016920423

About the Editor

Aldon Cris B. Galido

Aldon is currently taking up his MS EE (Microelectronics) degree at the University of the Philippines - Diliman Campus. His interests include low voltage and low power integrated circuits for wireless sensor node applications. He is currently working as a Research Fellow for the Resilient Sensory Swarms for Smart Energy and Environmental Monitoring (RESE2NSE) project at the Philippine-California Advanced Research Institute.

Table of Contents

List of Abbreviations xiii
Preface xv

1 FUZZY LOGIC

Introduction 1
Applying Truth Values 2
Linguistic Variables 3
Forming A Consensus of Inputs and Fuzzy Rules 3
Process 3
Early Applications 4
Example 4
Hard science with IF-THEN rules 4
Define with multiply 5
Define with Sigmoid 5
Propositional Fuzzy Logics 5
Predicate Fuzzy Logics 6
Decidability Issues for Fuzzy Logic 6
Fuzzy Databases 7
Comparison to Probability 7
Relation to Ecorithms 7
Compensatory Fuzzy Logic 8
Fuzzy Classification 8
Defuzzification 9
Methods 10
Expert System 11
Eliciting and Integrating Knowledge with Data 13
Software Architecture 13
Advantages 14
Disadvantages 15
Applications 16
References 17

2 FUZZY CONCEPT

Introduction 20
Origin and Etymology 20
Applications 21
Fuzzy Concept Lattices and Big Data 22
Definitional Controversies 24
Sociology and Journalism 25
Uncertainty 26
Language 27
Psychology 27
Everyday Occurrence 29
Economy of Distinctions 31
Analysis 32
References 34

3 FUZZY SET

Introduction 40
Fuzzy Logic 41
Fuzzy Number 41
Fuzzy Interval 41
Fuzzy Categories 41
Fuzzy Relation Equation 42
Entropy 42
Extensions 42
Fuzzy Set Operations 43
    Standard Fuzzy Set Operations 43
    Standard Complement 43
    Standard Intersection 44
    Standard Union 44
    Fuzzy Complements 44
    Fuzzy Intersections 44
    Fuzzy Unions 45
Rough Fuzzy Hybridization 46
    Type-2 Fuzzy Sets and Systems 46
    Interval Type-2 Fuzzy Sets 48
    Interval Type-2 Fuzzy Logic Systems 48
    Computing With Words 50
    Applications 51
Neuro-Fuzzy 51

Pseudo Outer-Product-Based Fuzzy Neural Networks 52
Fuzzy Control System.... 52
History and Applications....53
Building a Fuzzy Controller 60
Antilock Brakes 63
Logical Interpretation of Fuzzy Control....64
Fuzzy Finite Element 64
Parabolic Equation 64
References.... 65

4 CLASSICAL LOGIC

Introduction 71
Examples of Classical Logics....72
Generalized Semantics 72
Non-Classical Logics 72
Law of excluded middle 73
Classic Laws of Thought....73
Examples 74
Use in Computer Science Proofs....78
Criticisms....79
Double Negation 79
Double Negative Elimination....80
Law Of Noncontradiction 81
Interpretations 82
Principle of Explosion 86
Symbolic Representation 86
Arguments for Explosion....86
The Semantic Argument 87
Addressing the Principle 87
Use....87
Monotonicity of Entailment.... 87
Non-Monotonic Logics....88
Commutativity of Conjunction 88
Formal Notation 88
Generalized Principle 89
De Morgan's Laws.... 89
Engineering....91
Informal proof 92
Conclusion 94
Generalising De Morgan Duality....94
Extension to Predicate and Modal Logic....95

Propositional Calculus 95
Terminology 97
Closure under Operations 99
Argument 100
Generic Description of a Propositional Calculus 101
Example 1. Simple Axiom System 102
Example 2. Natural Deduction System 102
Negation Introduction 103
Negation Elimination 103
Double Negative Elimination 103
Conjunction Introduction 104
Conjunction Elimination 104
Disjunction Introduction 104
Disjunction Elimination 104
Biconditional Introduction 104
Biconditional Elimination 104
Modus Ponens (Conditional Elimination) 104
Conditional Proof (Conditional Introduction) 104
Basic and Derived Argument Forms 105
Proofs in Propositional Calculus 106
Example of a Proof 106
Soundness and Completeness of the Rules 106
QED 109
Another Outline for a Completeness Proof 109
Interpretation of a truth-Functional Propositional Calculus 109
Interpretation of a Sentence of Truth-Functional Propositional Logic 110
Alternative Calculus 110
Inference Rule 111
Meta-inference Rule 111
If the Sequence 111
Equivalence to Equational Logics 113
Graphical Calculi 114
Other Logical Calculi 114
Sorites Paradox 115
The Original Formulation and Variations 116
Proposed Resolutions 117
Dropping transitivity of the relations involved 119
References 120

5 DEGREE OF TRUTH

Half-Truth ... 124
- Purpose ... 124
- Examples ... 124
- Politics ... 125
- Meme Theory ... 125

Quotations ... 125

Truth ... 126
- Definition and Etymology ... 127
- Major Theories ... 128
- Substantive Theories ... 129
- Formal Theories ... 134
- Semantic Theory of Truth ... 135
- In Medicine and Psychiatry ... 144

Truth Value ... 144
- Intuitionistic and Constructive Logic ... 145
- Intuitionistic and Constructive Logic ... 145
- Multi-valued Logic ... 146
- Algebraic Semantics ... 146
- In other Theories ... 146
- Algebraic Semantics ... 146
- In other Theories ... 147

Vagueness ... 147
- Importance ... 147
- Approaches ... 148
- Legal Principle ... 149

Many-Valued Logic ... 149
- Examples ... 150
- Łukasiewicz logics Lv and L∞ ... 151
- Product logic Π ... 152
- Relation to Classical Logic ... 152

Three-Valued Logic ... 154
- Representation of Values ... 154
- Kleene and Priest Logics ... 155
- Łukasiewicz Logic ... 156
- Application in Sql ... 158

Four-Valued LogiC ... 158

Truth Function ... 159
- Table of Binary Truth Functions ... 159

Functional Completeness 163
One Element 164
Two Elements 164
Three Elements 164
Algebraic Properties 164
Principle of Compositionality 166
Computer Science 166
References 167

6 INTERVAL ARITHMETIC

Introduction 172
Example 172
Simple Arithmetic 173
Notation 175
Elementary Functions 176
Interval Extensions of General Functions 177
Complex Interval Arithmetic 178
Interval Methods 179
Rounded Interval Arithmetic 179
Dependency Problem 180
Linear Interval Systems 182
Interval Newton method 183
Bisection and Covrs 184
Application 185
Rounding Error Analysis 185
Tolerance Analysis 185
Fuzzy Interval Arithmetic 186
History 186
Implementations 187
Conferences And Workshop 188
References 189

7 FUZZY MEASURE THEORY

Introduction 191
Definitions 192
Simplification Assumptions for Fuzzy Measures 193
Sugeno λ-Measure 193
Definition 193
k-additive Fuzzy Measure 194
Example 195

Properties 196
Addendum Definitions 196
Aumann–Shapley Value....197
References....198

8 FIELD-PROGRAMMABLE GATE ARRAY

Introducton....199
Technical Design....201
History 201
Modern Developments....202
Gates....203
Market size 203
Design starts....203
Comparisons 204
Complex Programmable Logic Devices (CPLD) 204
Security Considerations 205
Applications 205
Architecture....208
Logic Blocks 208
Hard blocks 209
Clocking....209
3D Architectures....210
Design And Programming....210
Basic Process Technology Types 211
Major Manufacturers 211
Hardware Description Language....212
Motivation 212
library iEEE; 213
Structure of HDL....213
Design using HDL....215
Simulating and Debugging HDL code....216
Design Verification with HDLs 217
HDL and Programming Languages 217
High-level Synthesis 218
Examples of HDLs 218
Static Random-Access Memory 220
Characteristics 221
Types of SRAM 223
By flip-flop type....224
SRAM Operation 226
Flash Memory 227

Principles of Operation 229
Limitations 234
Low-level access 236
Flash file Systems 240
Applications 241
Flash memory as a replacement for hard drives 243
Industry 244
Flash Scalability 244
References 245

Index 253

List of Abbreviations

ADC	Analog-to-Digital Converters
ASIC	Application-Specific Integrated circuit
ATG	Automatic Test Pattern Generation
BMI	Body Mass Index
CLB	Configurable Logic Blocks
DAC	Digital-to-Analog Converters
DIC	Discrete Incremental Clustering
FKP	Fuzzy Kohonen Partitioning
FNN	Fuzzy Neural Network
FOU	Footprint of Uncertainty
FPGA	Field Programmable Gate Arrays
FPGA	Field-Programmable Gate Array
LMF	Lower Membership Function
LVQ	Learning Vector Quantization
NFS	Neuro-Fuzzy System
OWA	Ordered Weighted Averaging
PLA	Programmable Logic Arrays
PLD	Programmable Logic Devices
PROM	Programmable Read-Only Memory
SSD	Solid-State Drive
TSK	Takagi-Sugeno-Kang

Preface

Fuzzy logic is an approach to computing based on "degrees of truth" rather than the usual "true or false" (1 or 0) Boolean logic on which the modern computer is based. The idea of fuzzy logic was first advanced by Dr. Lotfi Zadeh of the University of California at Berkeley in the 1960s. Dr. Zadeh was working on the problem of computer understanding of natural language. Natural language (like most other activities in life and indeed the universe) is not easily translated into the absolute terms of 0 and 1. (Whether everything is ultimately describable in binary terms is a philosophical question worth pursuing, but in practice much data we might want to feed a computer is in some state in between and so, frequently, are the results of computing.) It may help to see fuzzy logic as the way reasoning really works and binary or Boolean logic is simply a special case of it. Fuzzy logic starts with and builds on a set of user-supplied human language rules. The fuzzy systems convert these rules to their mathematical equivalents. This simplifies the job of the system designer and the computer, and results in much more accurate representations of the way systems behave in the real world. Additional benefits of fuzzy logic include its simplicity and its flexibility. Fuzzy logic can handle problems with imprecise and incomplete data, and it can model nonlinear functions of arbitrary complexity. You can create a fuzzy system to match any set of input-output data. The Fuzzy Logic Toolbox makes this particularly easy by supplying adaptive techniques such as adaptive neuro-fuzzy inference systems (ANFIS) and fuzzy subtractive clustering. Fuzzy logic models, called fuzzy inference systems, consist of a number of conditional "if-then" rules. For the designer who understands the system, these rules are easy to write, and as many rules as necessary can be supplied to describe the system adequately (although typically only a moderate number of rules are needed). In fuzzy logic, unlike standard conditional logic, the truth of any statement is a matter of degree. (How cold is it? How high should we set the heat?) We are familiar with inference rules of the form p -> q (p implies q). With fuzzy logic, it's possible to say (.5* p) -> (.5 * q). For example, for the rule if (weather is cold) then (heat is on), both variables, cold and on, map to ranges of values. Fuzzy inference systems rely on membership functions to explain to the computer how to calculate the correct value between 0 and 1. The most tangible applications of fuzzy logic control have appeared commercial appliances. Specifically, but not limited to heating ventilation and air conditioning (HVAC) systems. These systems use fuzzy logic thermostats to control the heating and cooling, this saves energy by making the system more efficient.

Content Coverage

Chapters One and *Two* present the basics of fuzzy logic and fuzzy concept, respectively. A fuzzy concept is a concept of which the boundaries of application can vary considerably according to context or conditions, instead of being fixed once and for all.

Chapters Three and *Four* provide coverage of fuzzy sets and classical logic, separately. A fuzzy set operation is an operation on fuzzy sets. These operations are generalization of crisp set operations. Classical logic is an intensively studied and widely used class of formal logics.

Chapters Five and *Six* are degree of truth and interval arithmetic, correspondingly. Main focus of interval arithmetic is the simplest way to calculate upper and lower endpoints for the range of values of a function in one or more variables.

Chapters Seven and *Eight* present about fuzzy measure theory and field-programmable gate array (FPGA) technology, respectively. Fuzzy measure theory considers generalized measures in which the additive property is replaced by the weaker property of monotonicity.

CHAPTER 1

Fuzzy Logic

Chapter Contents

- **Introduction**
- **Forming A Consensus of Inputs and Fuzzy Rules**
- **Process**
- **Early Applications**
- **Example**
- **Fuzzy Databases**
- **Comparison to Probability**
- **Relation to Ecorithms**
- **Compensatory Fuzzy Logic**
- **Fuzzy Classification**
- **Defuzzification**
- **Expert System**

Fuzzy logic is a form of many-valued logic in which the truth values of variables may be any real number between 0 and 1. By contrast, in Boolean logic, the truth values of variables may only be the "crisp" values 0 or 1. Fuzzy logic has been employed to handle the concept of partial truth, where the truth value may range between completely true and completely false.Furthermore, when linguistic variables are used, these degrees may be managed by specific (membership) functions.

The term *fuzzy logic* was introduced with the 1965 proposal of fuzzy set theory by Lotfi Zadeh. Fuzzy logic had however been studied since the 1920s, as infinite-valued logic—notably by Łukasiewicz and Tarski. Fuzzy logic has been applied to many fields, from control theory to artificial intelligence.

INTRODUCTION

Classical logic only permits conclusions which are either true or false. However, there are also propositions with variable answers, such as one might find when asking a group of people to identify a colour. In such instances, the truth appears as the result of reasoning from inexact or partial knowledge in which the sampled answers are mapped on a spectrum.

Humans and animals often operate using fuzzy evaluations in many everyday situations. In the case where someone is tossing an object into a container from a distance, the person does not compute exact values for the object weight, density, distance, direction, container height and width, and air resistance to determine the force and angle to toss the object. Instead the person instinctively applies quick "fuzzy" estimates, based upon previous experience, to determine what output values of force, direction and vertical angle to use to make the toss.

Both degrees of truth and probabilities range between 0 and 1 and hence may seem similar at first, but fuzzy logic uses degrees of truth as a mathematical model of *vagueness*, while probability is a mathematical model of *ignorance*.

Take, for example, the concepts of "empty" and "full". The meaning of each of them can be represented by a certain fuzzy set. The concept of emptiness would be subjective and thus would depend on the observer or designer. A 100 ml glass containing 30 ml of water may be defined as being 0.7 empty and 0.3 full, but another designer might, equally well, design a set membership function where the glass would be considered full for all values down to 50 ml.

Applying Truth Values

A basic application might characterize various sub-ranges of a continuous variable. For instance, a temperature measurement for anti-lock brakes might have several separate membership functions defining particular temperature ranges needed to control the brakes properly. Each function maps the same temperature value to a truth value in the 0 to 1 range. These truth values can then be used to determine how the brakes should be controlled.

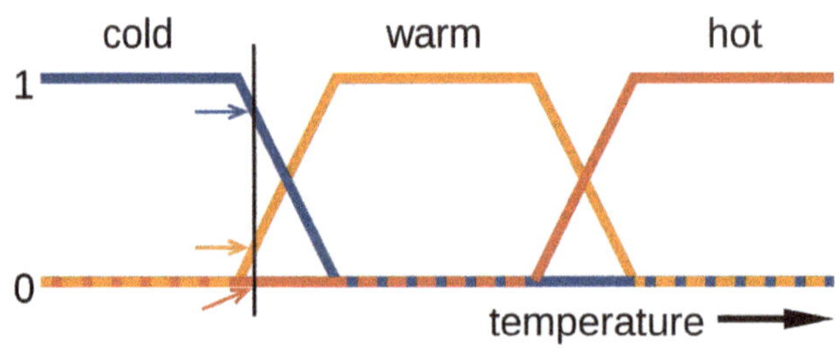

Fuzzy logic temperature

In this image, the meanings of the expressions *cold*, *warm*, and *hot* are represented by functions mapping a temperature scale. A point on that scale has three "truth values"—one for each of the three functions. The vertical line in the image represents a particular temperature that the three arrows (truth values) gauge. Since the red arrow points to zero, this temperature may be interpreted as "not hot". The orange arrow (pointing at 0.2) may describe it as "slightly warm" and the blue arrow (pointing at 0.8) "fairly cold".

Linguistic Variables

While variables in mathematics usually take numerical values, in fuzzy logic applications non-numeric values are often used to facilitate the expression of rules and facts.

A linguistic variable such as *age* may accept values such as *young* and its antonym *old*. Because natural languages do not always contain enough value terms to express a fuzzy value scale, it is common practice to modify linguistic values with adjectives or adverbs. For example, we can use the hedges *rather* and *somewhat* to construct the additional values *rather old* or *somewhat young*. Fuzzification operations can map mathematical input values into fuzzy membership functions. And the opposite de-fuzzifying operations can be used to map a fuzzy output membership functions into a "crisp" output value that can be then used for decision or control purposes.

FORMING A CONSENSUS OF INPUTS AND FUZZY RULES

Since the fuzzy system output is a consensus of all of the inputs and all of the rules, fuzzy logic systems can be well behaved when input values are not available or are not trustworthy. Weightings can be optionally added to each rule in the rulebase and weightings can be used to regulate the degree to which a rule affects the output values. These rule weightings can be based upon the priority, reliability or consistency of each rule. These rule weightings may be static or can be changed dynamically, even based upon the output from other rules.

PROCESS

- Fuzzify all input values into fuzzy membership functions.
- Execute all applicable rules in the rulebase to compute the fuzzy output functions.
- De-fuzzify the fuzzy output functions to get "crisp" output values.

Synthesis of fuzzy logic functions given in tabular form

Fuzzy logic function of Zadeh takes value of one of its arguments or a negation of an argument. Thus, a fuzzy logic function can be given by a choice table where all the variants of ordering of arguments and their negations are listed. For example, a row of a choice table of a two arguments function can look as follows:

$$x_1 \le x_2 \le \bar{x}_2 \le \bar{x}_1 : x_2$$

However, an arbitrary choice table not always defines a fuzzy logic function. In the paper, a criterion has been formulated to recognize whether a given choice table defines a fuzzy logic function and a simple algorithm of fuzzy logic function synthesis has been proposed based on introduced concepts of constituents of minimum and maximum. A fuzzy logic function represents a disdjunction of constituents of minimum, where a constituent of minimum is a conjunction of variables of the current area greater than or equal to the function value in this area (to the right of the function value in the inequality, including the function value). For the above example row, the constituent of minimum is $x_2\bar{x}_2\bar{x}_1$.

EARLY APPLICATIONS

Many of the early successful applications of fuzzy logic were implemented in Japan. The first notable application was on the high-speed train in Sendai, in which fuzzy logic was able to improve the economy, comfort, and precision of the ride. It has also been used in recognition of hand written symbols in Sony pocket computers, flight aid for helicopters, controlling of subway systems in order to improve driving comfort, precision of halting, and power economy, improved fuel consumption for automobiles, single-button control for washing machines, automatic motor control for vacuum cleaners with recognition of surface condition and degree of soiling, and prediction systems for early recognition of earthquakes through the Institute of Seismology Bureau of Meteorology, Japan.

EXAMPLE

Hard science with IF-THEN rules

Fuzzy set theory defines fuzzy operators on fuzzy sets. The problem in applying this is that the appropriate fuzzy operator may not be known. For example, the logic for a simple temperature regulator that uses a fan might look like this:

IF temperature IS very cold THEN stop fan

IF temperature IS cold THEN fan speed is slow

IF temperature IS warm THEN fan speed is moderate

IF temperature IS hot THEN fan speed is high

Using this rulebase and the previous image, we would expect the output fan speed to be a combination of zero and moderate, which would evaluated as some degree of slow when the input value is a combination of cold and warm and not hot. The fan speed will continue to get slower as the input temperature gets colder until the input temperature is 100% cold, 0% warm and 0% hot, at which point the output fan speed will be zero. As the temperature input gets warmer and hotter, the output fan speed will continue to get faster until the input temperature is 0% cold, 0% warm and 100% hot, at which point the fan speed output will be high.

If the fuzzy membership functions cover 100% of the input variable domain, then it can be proven that the behavior of the fuzzy system is fully deterministic over the entire input domain and nowhere ambiguous. This determinism is very important for use in control and decision systems.

There is no "ELSE"—all of the rules are evaluated, because the temperature might be "cold" and "normal" at the same time to different degrees.

The AND, OR, and NOT operators of Boolean logic exist in fuzzy logic, usually defined as the minimum, maximum, and complement; when they are defined this way, they are called the *Zadeh operators*. So for the fuzzy variables x and y:

NOT x = (1 - truth(x))

x AND y = minimum(truth(x), truth(y))

x OR y = maximum(truth(x), truth(y))

There are also other operators, more linguistic in nature, called *hedges* that can be applied. These are generally adverbs such as *very*, or *somewhat*, which modify the meaning of a set using a mathematical formula.

Define with multiply

x AND y = x*y

x OR y = 1-(1-x)*(1-y) = x+y-x*y

1-(1-x)*(1-y) comes from this:

x OR y = NOT(AND(NOT(x), NOT(y)))

x OR y = NOT(AND(1-x, 1-y))

x OR y = NOT((1-x)*(1-y))

x OR y = 1-(1-x)*(1-y)

Define with Sigmoid

Fuzzy sets can be defined using the sigmoid function.

sigmoid(x)=1/(1+e^-x)

sigmoid(x)+sigmoid(-x) = 1

(sigmoid(x)+sigmoid(-x))*(sigmoid(y)+sigmoid(-y))*(sigmoid(z)+sigmoid(-z)) = 1

Logical analysis

In mathematical logic, there are several formal systems of "fuzzy logic"; most of them belong among so-called t-norm fuzzy logic.

Propositional Fuzzy Logics

The most important propositional fuzzy logics are:-

- Monoidal t-norm-based propositional fuzzy logic MTL is an axiomatization of logic where conjunction is defined by a left continuous t-norm and implication is defined as the residuum of the t-norm. Its models correspond to MTL-algebras that are pre-linear commutative bounded integral residuated lattices.
- Basic propositional fuzzy logic BL is an extension of MTL logic where conjunction is defined by a continuous t-norm, and implication is also defined as the residuum of the t-norm. Its models correspond to BL-algebras.
- Łukasiewicz fuzzy logic is the extension of basic fuzzy logic BL where standard conjunction is the Łukasiewicz t-norm. It has the axioms of basic fuzzy logic plus an axiom of double negation, and its models correspond to MV-algebras.

- Gödel fuzzy logic is the extension of basic fuzzy logic BL where conjunction is Gödel t-norm. It has the axioms of BL plus an axiom of idempotence of conjunction, and its models are called G-algebras.
- Product fuzzy logic is the extension of basic fuzzy logic BL where conjunction is product t-norm. It has the axioms of BL plus another axiom for cancellativity of conjunction, and its models are called product algebras.
- Fuzzy logic with evaluated syntax (sometimes also called Pavelka"s logic), denoted by EVŁ, is a further generalization of mathematical fuzzy logic. While the above kinds of fuzzy logic have traditional syntax and many-valued semantics, in EVŁ is evaluated also syntax. This means that each formula has an evaluation. Axiomatization of EVŁ stems from Łukasziewicz fuzzy logic. A generalization of classical Gödel completeness theorem is provable in EVŁ.

Predicate Fuzzy Logics

These extend the above-mentioned fuzzy logics by adding universal and existential quantifiers in a manner similar to the way that predicate logic is created from propositional logic. The semantics of the universal (resp. existential) quantifier in t-norm fuzzy logics is the infimum (resp. supremum) of the truth degrees of the instances of the quantified subformula.

Decidability Issues for Fuzzy Logic

The notions of a "decidable subset" and "recursively enumerable subset" are basic ones for classical mathematics and classical logic. Thus the question of a suitable extension of them to fuzzy set theory is a crucial one. A first proposal in such a direction was made by E.S. Santos by the notions of *fuzzy Turing machine*, *Markov normal fuzzy algorithm* and *fuzzy program* (see Santos 1970). Successively, L. Biacino and G. Gerla argued that the proposed definitions are rather questionable. For example, in one shows that the fuzzy Turing machines are not adequate for fuzzy language theory since there are natural fuzzy languages intuitively computable that cannot be recognized by a fuzzy Turing Machine. Then, they proposed the following definitions. Denote by *Ü* the set of rational numbers in [0,1]. Then a fuzzy subset s : $S \to [0,1]$ of a set S is recursively enumerable if a recursive map h : $S \times N \to Ü$ exists such that, for every x in S, the function $h(x,n)$ is increasing with respect to n and $s(x) = \lim h(x,n)$. We say that s is *decidable* if both s and its complement $-s$ are recursively enumerable. An extension of such a theory to the general case of the L-subsets is possible (see Gerla 2006). The proposed definitions are well related with fuzzy logic. Indeed, the following theorem holds true (provided that the deduction apparatus of the considered fuzzy logic satisfies some obvious effectiveness property). **Theorem.** Any axiomatizable fuzzy theory is recursively enumerable. In particular, the fuzzy set of logically true formulas is recursively enumerable in spite of the fact that the crisp set of valid formulas is not recursively enumerable, in general. Moreover, any axiomatizable and complete theory is decidable.

It is an open question to give supports for a "Church thesis" for fuzzy mathematics, the proposed notion of recursive enumerability for fuzzy subsets is the adequate one. To this aim, an extension of the notions of fuzzy grammar and fuzzy Turing machine should be necessary. Another open question is to start from this notion to find an extension of Gödel"s theorems to fuzzy logic.

FUZZY DATABASES

Once fuzzy relations are defined, it is possible to develop fuzzy relational databases. The first fuzzy relational database, FRDB, appeared in Maria Zemankova's dissertation (1983). Later, some other models arose like the Buckles-Petry model, the Prade-Testemale Model, the Umano-Fukami model or the GEFRED model by J.M. Medina, M.A. Vila et al.

Fuzzy querying languages have been defined, such as the SQLf by P. Bosc et al. and the FSQL by J. Galindo et al. These languages define some structures in order to include fuzzy aspects in the SQL statements, like fuzzy conditions, fuzzy comparators, fuzzy constants, fuzzy constraints, fuzzy thresholds, linguistic labels etc.

COMPARISON TO PROBABILITY

Fuzzy logic and probability address different forms of uncertainty. While both fuzzy logic and probability theory can represent degrees of certain kinds of subjective belief, fuzzy set theory uses the concept of fuzzy set membership, i.e., *how much* a variable is in a set (there is not necessarily any uncertainty about this degree), and probability theory uses the concept of subjective probability, i.e., *how probable* is it that a variable is in a set (it either entirely is or entirely is not in the set in reality, but there is uncertainty around whether it is or is not). The technical consequence of this distinction is that fuzzy set theory relaxes the axioms of classical probability, which are themselves derived from adding uncertainty, but not degree, to the crisp true/false distinctions of classical Aristotelian logic. Bruno de Finetti controversially argues that only one kind of mathematical uncertainty, probability, is needed, and thus fuzzy logic is superfluous. However, Bart Kosko shows in Fuzziness vs. Probability that probability theory is a subtheory of fuzzy logic, as questions of degrees of belief in mutually-exclusive set membership in probability theory can be represented as certain cases of non-mutually-exclusive graded membership in fuzzy theory. In that context, he also derives Bayes" theorem from the concept of fuzzy subsethood. Lotfi A. Zadeh argues that fuzzy logic is different in character from probability, and is not a replacement for it. He fuzzified probability to fuzzy probability and also generalized it to possibility theory. (cf.)

More generally, fuzzy logic is one of many different extensions to classical logic intended to deal with issues of uncertainty outside of the scope of classical logic, the inapplicability of probability theory in many domains, and the paradoxes of Dempster-Shafer theory.

RELATION TO ECORITHMS

Computational theorist Leslie Valiant uses the term *ecorithms* to describe how many less exact systems and techniques like fuzzy logic (and "less robust" logic) can be applied to learning algorithms. Valiant essentially redefines machine learning as evolutionary. In general use, ecorithms are algorithms that learn from their more complex environments (hence *eco-*) to generalize, approximate and simplify solution logic. Like fuzzy logic, they are methods used to overcome continuous variables or systems too complex to completely enumerate or understand discretely or exactly. Ecorithms and fuzzy logic also have the common property of dealing with possibilities more than probabilities, although feedback and feed forward, basically stochastic weights, are a feature of both when dealing with, for example, dynamical systems.

COMPENSATORY FUZZY LOGIC

Compensatory fuzzy logic (CFL) is a branch of fuzzy logic with modified rules for conjunction and disjunction. When the truth value of one component of a conjunction or disjunction is increased or decreased, the other component is decreased or increased to compensate. This increase or decrease in truth value may be offset by the increase or decrease in another component. An offset may be blocked when certain thresholds are met. Proponents claim that CFL allows for better computational semantic behaviors. http://web.mit.edu/6.863/www/fall2012/projects/writeups/semantic-similarity-betweenverbs.pdf

Compensatory Fuzzy Logic consists of four continuous operators: conjunction (c); disjunction (d); fuzzy strict order (or); and negation (n). The conjunction is the geometric mean and its dual as conjunctive and disjunctive operators.

FUZZY CLASSIFICATION

Fuzzy classification is the process of grouping elements into a fuzzy set whose membership function is defined by the truth value of a fuzzy propositional function.

A fuzzy class ~C = { i | ~Π(i) } is defined as a fuzzy set ~C of individuals i satisfying a fuzzy classification predicate ~Π which is a fuzzy propositional function. The domain of the fuzzy class operator ~{ .| .} is the set of variables V and the set of fuzzy propositional functions ~PF, and the range is the fuzzy powerset (the set of fuzzy subsets) of this universe, ~P(U):

~{ .| .}:V × ~PF ⟶ ~P(U)

A fuzzy propositional function is, analogous to, an expression containing one or more variables, such that, when values are assigned to these variables, the expression becomes a fuzzy proposition in the sense of. Accordingly, fuzzy classification is the process of grouping individuals having the same characteristics into a *fuzzy set*. A fuzzy classification corresponds to a membership function μ that indicates whether an individual is a member of a class, given its fuzzy classification predicate ~Π.

μ:~PF × U ⟶ ~T

Here, ~T is the set of fuzzy truth values (the interval between zero and one). The fuzzy classification predicate ~Π corresponds to a fuzzy restriction "i is R" of U, where R is a fuzzy set defined by a truth function. The degree of membership of an individual i in the fuzzy class ~C is defined by the truth value of the corresponding fuzzy predicate.

μ~C(i):= τ(~Π(i))

Intuitively, a class is a set that is defined by a certain property, and all objects having that property are elements of that class. The process of classification evaluates for a given set of objects whether they fulfill the classification property, and consequentially are a member of the corresponding class. However, this intuitive concept has some logical subtleties that need clarification.

A class logic is a logical system which supports set construction using logical predicates with the class operator $\{ .|.\}$. A *class*

$C = \{ i \mid \Pi(i) \}$

is defined as a set C of individuals i satisfying a classification predicate Π which is a propositional function. The domain of the class operator $\{ .|.\}$ is the set of variables V and the set of propositional functions PF, and the range is the powerset of this universe P(U) that is, the set of possible subsets:

$\{ .|.\} : V \times PF \longrightarrow P(U)$

Here is an explanation of the logical elements that constitute this definition:

- An individual is a real object of reference.
- A universe of discourse is the set of all possible individuals considered.
- A variable $V{:}\longrightarrow R$ is a function which maps into a predefined range R without any given function arguments: a zero-place function.
- A propositional function is "an expression containing one or more undetermined constituents, such that, when values are assigned to these constituents, the expression becomes a proposition".

In contrast, *classification* is the process of grouping individuals having the same characteristics into a set. A classification corresponds to a membership function μ that indicates whether an individual is a member of a class, given its classification predicate Π.

$\mu{:}PF \times U \longrightarrow T$

The membership function maps from the set of propositional functions PF and the universe of discourse U into the set of truth values T. The membership μ of individual i in Class C is defined by the truth value τ of the classification predicate Π.

$\mu C(i) := \tau(\Pi(i))$

In classical logic the truth values are certain. Therefore a classification is crisp, since the truth values are either exactly true or exactly false.

DEFUZZIFICATION

Defuzzification is the process of producing a quantifiable result in Crisp logic, given fuzzy sets and corresponding membership degrees. It is typically needed in fuzzy control systems. These will have a number of rules that transform a number of variables into a fuzzy result, that is, the result is described in terms of membership in fuzzy sets. For example, rules designed to decide how much pressure to apply might result in "Decrease Pressure (15%), Maintain Pressure (34%), Increase Pressure (72%)". Defuzzification is interpreting the membership degrees of the fuzzy sets into a specific decision or real value.

The simplest but least useful defuzzification method is to choose the set with the highest membership, in this case, "Increase Pressure" since it has a 72% membership, and ignore the others, and convert this 72% to some number. The problem with this approach is that it loses information. The rules that called for decreasing or maintaining pressure might as well have not been there in this case.

A common and useful defuzzification technique is *center of gravity*. First, the results of the rules must be added together in some way. The most typical fuzzy set membership function has the graph of a triangle. Now, if this triangle were to be cut in a straight horizontal line somewhere between the top and the bottom, and the top portion were to be removed, the remaining portion forms a trapezoid. The first step of defuzzification typically "chops off" parts of the graphs to form trapezoids (or other shapes if the initial shapes were not triangles). For example, if the output has "Decrease Pressure (15%)", then this triangle will be cut 15% the way up from the bottom. In the most common technique, all of these trapezoids are then superimposed one upon another, forming a single geometric shape. Then, the centroid of this shape, called the *fuzzy centroid*, is calculated. The *x* coordinate of the centroid is the defuzzified value.

Methods

There are many different methods of defuzzification available, including the following:

- AI (adaptive integration)
- BADD (basic defuzzification distributions)
- BOA (bisector of area)
- CDD (constraint decision defuzzification)
- COA (center of area)
- COG (center of gravity)
- ECOA (extended center of area)
- EQM (extended quality method)
- FCD (fuzzy clustering defuzzification)
- FM (fuzzy mean)
- FOM (first of maximum)
- GLSD (generalized level set defuzzification)
- ICOG (indexed center of gravity)
- IV (influence value)
- LOM (last of maximum)
- MeOM (mean of maxima)
- MOM (middle of maximum)
- QM (quality method)
- RCOM (random choice of maximum)
- SLIDE (semi-linear defuzzification)
- WFM (weighted fuzzy mean)

The maxima methods are good candidates for fuzzy reasoning systems. The distribution methods and the area methods exhibit the property of continuity that makes them suitable for fuzzy controllers.

EXPERT SYSTEM

In artificial intelligence, an **expert system** is a computer system that emulates the decision-making ability of a human expert. Expert systems are designed to solve complex problems by reasoning about knowledge, represented mainly as if–then rules rather than through conventional procedural code. The first expert systems were created in the 1970s and then proliferated in the 1980s. Expert systems were among the first truly successful forms of artificial intelligence (AI) software.

A Symbolics Lisp Machine: an early platform for expert systems. Note the unusual *space-cadet keyboard.*

An expert system is divided into two subsystems: the inference engine and the knowledge base. The knowledge base represents facts and rules. The inference engine applies the rules to the known facts to deduce new facts. Inference engines can also include explanation and debugging abilities.

Expert systems were introduced by the Stanford Heuristic Programming Project led by Edward Feigenbaum, who is sometimes termed the "father of expert systems"; other key early contributors were Jairus Lainibo, Bruce Buchanan, and Randall Davis. The Stanford researchers tried to identify domains where expertise was highly valued and complex, such as diagnosing infectious diseases (Mycin) and identifying unknown organic molecules (Dendral). Although that "intelligent systems derive their power from the knowledge they possess rather than from the specific formalisms and inference schemes they use" – as Feigenbaum said – seems in retrospect a rather straightforward insight, it was a significant step forward then, since until then, research had been focused on attempts to develop very general-purpose problem solvers, such as those described by Allen

Newell and Herb Simon. Expert systems became some of the first truly successful forms of artificial intelligence (AI) software.

Research on expert systems was also active in France. While in the US the focus tended to be on rule-based systems, first on systems hard coded on top of LISP programming environments and then on expert system shells developed by vendors such as Intellicorp, in France research focused more on systems developed in Prolog. The advantage of expert system shells was that they were somewhat easier for nonprogrammers to use. The advantage of Prolog environments was that they weren"t focused only on *if-then* rules; Prolog environments provided a much fuller realization of a complete First Order Logic environment.

In the 1980s, expert systems proliferated. Universities offered expert system courses and two thirds of the Fortune 500 companies applied the technology in daily business activities. Interest was international with the Fifth Generation Computer Systems project in Japan and increased research funding in Europe.

In 1981, the first IBM PC, with the PC DOS operating system, was introduced. The imbalance between the high affordability of the relatively powerful chips in the PC, compared to the much more expensive cost of processing power in the mainframes that dominated the corporate IT world at the time, created a new type of architecture for corporate computing, termed the client-server model. Calculations and reasoning could be performed at a fraction of the price of a mainframe using a PC. This model also enabled business units to bypass corporate IT departments and directly build their own applications. As a result, client server had a tremendous impact on the expert systems market. Expert systems were already outliers in much of the business world, requiring new skills that many IT departments did not have and were not eager to develop. They were a natural fit for new PC-based shells that promised to put application development into the hands of end users and experts. Until then, the main development environment for expert systems had been high end Lisp machines from Xerox, Symbolics, and Texas Instruments. With the rise of the PC and client server computing, vendors such as Intellicorp and Inference Corporation shifted their priorities to developing PC based tools. Also, new vendors, often financed by venture capital (such as Aion Corporation, Neuron Data, Exsys, and many others), started appearing regularly.

In the 1990s and beyond, the term *expert system* and the idea of a standalone AI system mostly dropped from the IT lexicon. There are two interpretations of this. One is that "expert systems failed": the IT world moved on because expert systems didn't deliver on their over hyped promise. The other is the mirror opposite, that expert systems were simply victims of their success: as IT professionals grasped concepts such as rule engines, such tools migrated from being standalone tools for developing special purpose *expert* systems, to being one of many standard tools. Many of the leading major business application suite vendors (such as SAP, Siebel, and Oracle) integrated expert system abilities into their suite of products as a way of specifying business logic – rule engines are no longer simply for defining the rules an expert would use but for any type of complex, volatile, and critical business logic; they often go hand in hand with business process automation and integration environments.

Eliciting and Integrating Knowledge with Data

When developing expert systems it is common to integrate knowledge with data. Constantinou et al. proposed a method for eliciting and incorporating expert knowledge in data-driven Bayesian Networks (BNs). The method addresses the problem whereby the distribution of some variable in a BN is known from data, but where we wish to explicitly model the impact of some additional knowledge-based variables for which there is knowledge but no data. The method ensures that the expected value of a data-driven variable is preserved when a knowledge-based variable is incorporated as a cause (i.e. ancestor/parent) of the data-driven variable, as long as the knowledge-based variable remains unobserved. This is because the statistical outcomes of a data-driven variable are already influenced by the causes an expert might identify as variables missing form the dataset.

Software Architecture

An expert system is an example of a knowledge-based system. Expert systems were the first commercial systems to use a knowledge-based architecture. A knowledge-based system is essentially composed of two sub-systems: the knowledge base and the inference engine.

The knowledge base represents facts about the world. In early expert systems such as Mycin and Dendral, these facts were represented mainly as flat assertions about variables. In later expert systems developed with commercial shells, the knowledge base took on more structure and used concepts from object-oriented programming. The world was represented as classes, subclasses, and instances and assertions were replaced by values of object instances. The rules worked by querying and asserting values of the objects.

The inference engine is an automated reasoning system that evaluates the current state of the knowledge-base, applies relevant rules, and then asserts new knowledge into the knowledge base. The inference engine may also include abilities for explanation, so that it can explain to a user the chain of reasoning used to arrive at a particular conclusion by tracing back over the firing of rules that resulted in the assertion.

There are mainly two modes for an inference engine: forward chaining and backward chaining. The different approaches are dictated by whether the inference engine is being driven by the antecedent (left hand side) or the consequent (right hand side) of the rule. In forward chaining an antecedent fires and asserts the consequent. For example, consider the following rule:

$$R1 : Man(x) => Mortal(x)$$

A simple example of forward chaining would be to assert Man(Socrates) to the system and then trigger the inference engine. It would match R1 and assert Mortal(Socrates) into the knowledge base.

Backward chaining is a bit less straight forward. In backward chaining the system looks at possible conclusions and works backward to see if they might be true. So if the system was trying to determine if Mortal(Socrates) is true it would find R1 and query the knowledge base to see if Man(Socrates) is true. One of the early innovations of expert systems shells was to integrate inference engines with a user interface. This could be especially powerful with backward chaining. If the system needs to know a particular fact but doesn't it can simply generate an input screen and ask the user if the information is known. So in this example, it could use R1 to ask the user if Socrates was a Man and then use that new information accordingly.

The use of rules to explicitly represent knowledge also enabled explanation abilities. In the simple example above if the system had used R1 to assert that Socrates was Mortal and a user wished to understand why Socrates was mortal they could query the system and the system would look back at the rules which fired to cause the assertion and present those rules to the user as an explanation. In English if the user asked "Why is Socrates Mortal?" the system would reply "Because all men are mortal and Socrates is a man". A significant area for research was the generation of explanations from the knowledge base in natural English rather than simply by showing the more formal but less intuitive rules.

As expert systems evolved, many new techniques were incorporated into various types of inference engines. Some of the most important of these were:

- Truth maintenance. These systems record the dependencies in a knowledge-base so that when facts are altered, dependent knowledge can be altered accordingly. For example, if the system learns that Socrates is no longer known to be a man it will revoke the assertion that Socrates is mortal.
- Hypothetical reasoning. In this, the knowledge base can be divided up into many possible views, a.k.a. worlds. This allows the inference engine to explore multiple possibilities in parallel. For example, the system may want to explore the consequences of both assertions, what will be true if Socrates is a Man and what will be true if he is not?
- Fuzzy logic. One of the first extensions of simply using rules to represent knowledge was also to associate a probability with each rule. So, not to assert that Socrates is mortal, but to assert Socrates *may* be mortal with some probability value. Simple probabilities were extended in some systems with sophisticated mechanisms for uncertain reasoning and combination of probabilities.
- Ontology classification. With the addition of object classes to the knowledge base, a new type of reasoning was possible. Along with reasoning simply about object values, the system could also reason about object structures. In this simple example, Man can represent an object class and R1 can be redefined as a rule that defines the class of all men. These types of special purpose inference engines are termed classifiers. Although they were not highly used in expert systems, classifiers are very powerful for unstructured volatile domains, and are a key technology for the Internet and the emerging Semantic Web.

Advantages

The goal of knowledge-based systems is to make the critical information required for the system to work explicit rather than implicit. In a traditional computer program the logic is embedded in code that can typically only be reviewed by an IT specialist. With an expert system the goal was to specify the rules in a format that was intuitive and easily understood, reviewed, and even edited by domain experts rather than IT experts. The benefits of this explicit knowledge representation were rapid development and ease of maintenance.

Ease of maintenance is the most obvious benefit. This was achieved in two ways. First, by removing the need to write conventional code, many of the normal problems that can be caused by even small changes to a system could be avoided with expert systems. Essentially, the logical flow of the program (at least at the highest level) was simply a given for the system, simply invoke the inference engine. This also was a reason for the second benefit: rapid prototyping. With an expert system shell it was possible to enter a few rules and have a prototype developed in days rather than the months or year typically asociated with complex IT projects.

A claim for expert system shells that was often made was that they removed the need for trained programmers and that experts could develop systems themselves. In reality, this was seldom if ever true. While the rules for an expert system were more comprehensible than typical computer code, they still had a formal syntax where a misplaced comma or other character could cause havoc as with any other computer language. Also, as expert systems moved from prototypes in the lab to deployment in the business world, issues of integration and maintenance became far more critical. Inevitably demands to integrate with, and take advantage of, large legacy databases and systems arose. To accomplish this, integration required the same skills as any other type of system.

Disadvantages

The most common disadvantage cited for expert systems in the academic literature is the knowledge acquisition problem. Obtaining the time of domain experts for any software application is always difficult, but for expert systems it was especially difficult because the experts were by definition highly valued and in constant demand by the organization. As a result of this problem, a great deal of research in the later years of expert systems was focused on tools for knowledge acquisition, to help automate the process of designing, debugging, and maintaining rules defined by experts. However, when looking at the life-cycle of expert systems in actual use, other problems – essentially the same problems as those of any other large system – seem at least as critical as knowledge acquisition: integration, access to large databases, and performance.

Performance was especially problematic because early expert systems were built using tools such as Lisp, which executed interpreted (rather than compiled) code. Interpreting provided an extremely powerful development environment but with the drawback that it was virtually impossible to match the efficiency of the fastest compiled languages, such as C. System and database integration were difficult for early expert systems because the tools were mostly in languages and platforms that were neither familiar to nor welcome in most corporate IT environments – programming languages such as Lisp and Prolog, and hardware platforms such as Lisp machines and personal computers. As a result, much effort in the later stages of expert system tool development was focused on integrating with legacy environments such as COBOL and large database systems, and on porting to more standard platforms. These issues were resolved mainly by the client-server paradigm shift, as PCs were gradually accepted in the IT environment as a legitimate platform for serious business system development and as affordable minicomputer servers provided the processing power needed for AI applications.

Applications

Hayes-Roth divides expert systems applications into 10 categories illustrated in the following table. The example applications were not in the original Hayes-Roth table, and some of them arose well afterward. Any application that is not footnoted is described in the Hayes-Roth book. Also, while these categories provide an intuitive framework to describe the space of expert systems applications, they are not rigid categories, and in some cases an application may show traits of more than one category.

Category	Problem addressed	Examples
Interpretation	Inferring situation descriptions from sensor data	Hearsay (speech recognition), PROSPECTOR
Prediction	Inferring likely consequences of given situations	Preterm Birth Risk Assessment
Diagnosis	Inferring system malfunctions from observables	CADUCEUS, MYCIN, PUFF, Mistral, Eydenet,Kaleidos
Design	Configuring objects under constraints	Dendral, Mortgage Loan Advisor, R1 (DEC VAX Configuration)
Planning	Designing actions	Mission Planning for Autonomous Underwater Vehicle
Monitoring	Comparing observations to plan vulnerabilities	REACTOR
Debugging	Providing incremental solutions for complex problems	SAINT, MATHLAB, MACSYMA
Repair	Executing a plan to administer a prescribed remedy	Toxic Spill Crisis Management
Instruction	Diagnosing, assessing, and repairing student behavior	SMH.PAL, Intelligent Clinical Training,STEAMER
Control	Interpreting, predicting, repairing, and monitoring system behaviors	Real Time Process Control, Space Shuttle Mission Control

Hearsay was an early attempt at solving voice recognition through an expert systems approach. For the most part this category or expert systems was not all that successful. Hearsay and all interpretation systems are essentially pattern recognition systems—looking for patterns in noisy data. In the case of Hearsay recognizing phonemes in an audio stream. Other early examples were analyzing sonar data to detect Russian submarines. These kinds of systems proved much more amenable to a neural network AI solution than a rule-based approach.

CADUCEUS and MYCIN were medical diagnosis systems. The user describes their symptoms to the computer as they would to a doctor and the computer returns a medical diagnosis.

Dendral was a tool to study hypothesis formation in the identification of organic molecules. The general problem it solved—designing a solution given a set of constraints—was one of the most successful areas for early expert systems applied to business domains such as salespeople configuring Digital Equipment Corporation (DEC) VAX computers and mortgage loan application development.

SMH.PAL is an expert system for the assessment of students with multiple disabilities.

Mistral is an expert system to monitor dam safety, developed in the 90"s by Ismes (Italy). It gets data from an automatic monitoring system and performs a diagnosis of the state of the dam. Its first copy, installed in 1992 on the Ridracoli Dam (Italy), is still operational 24/7/365. It has been installed on several dams in Italy and abroad (e.g., Itaipu Dam in Brazil), and on landslide sites under the name of Eydenet, and on monuments under the name of Kaleidos. Mistral is a registered trade mark of CESI.

Applications of Expert Systems as Bayesian Networks

Bayesian Networks (BNs) are probabilistic graphical models, which are typically used to model cause and effect relationships, have become the most widely accepted technique for incorporating expert knowledge along with data. Expert knowledge can be incorporated into BNs by either constructing the causal (or dependence) graph, or by incorporating factors into the causal network which are important for inference but which data fail to capture. The popularity of BNs as expert systems has lead to the development of countless prediction and decision support systems in industry, government and academia worldwide. These systems typically incorporate both knowledge and data, and have been applied in the areas of, but not limited to, finance, engineering, sports , sports psychology, law, project management, marketing, medicine, energy, forensics, economics, property market, and defence.

REFERENCES

1. Novák, V., Perfilieva, I. and Močkoř, J. (1999) Mathematical principles of fuzzy logic Dodrecht: Kluwer Academic. ISBN 0-7923-8595-0
2. Ahlawat, Nishant, Ashu Gautam, and Nidhi Sharma (International Research Publications House 2014) "Use of Logic Gates to Make Edge Avoider Robot." International Journal of Information & Computation Technology (Volume 4, Issue 6; page 630) ISSN 0974-2239 (Retrieved 27 April 2014)
3. «Fuzzy Logic". Stanford Encyclopedia of Philosophy. Bryant University. 2006-07-23. Retrieved 2008-09-30.
4. Zadeh, L.A. (1965). "Fuzzy sets". Information and Control. 8 *(3): 338–353.* doi:10.1016/s0019-9958(65)90241-x.
5. Pelletier, Francis Jeffry (2000). "Review of Metamathematics of fuzzy logics" (PDF). The Bulletin of Symbolic Logic. 6 *(3): 342–346. JSTOR 421060.*
6. Zadeh, L. A. et al. 1996 Fuzzy Sets, Fuzzy Logic, Fuzzy Systems, World Scientific Press, ISBN 981-02-2421-4

7. Zaitsev, D.A.; Sarbei, V.G.; Sleptsov, A.I. (1998). "Synthesis of continuous-valued logic functions defined in tabular form". Cybernetics and Systems Analysis. 34 *(2): 190–195.* doi:10.1007/BF02742068.

8. Kosko, B (June 1, 1994). "Fuzzy Thinking: The New Science of Fuzzy Logic". Hyperion.

9. Bansod, Nitin A., Marshall Kulkarni, and S.H. Patil (Bharati Vidyapeeth College of Engineering) "Soft Computing- A Fuzzy Logic Approach". Soft Computing (Allied Publishers 2005) (page 73)

10. Wierman, Mark J. "An Introduction to the Mathematics of Uncertainty: including Set Theory, Logic, Probability, Fuzzy Sets, Rough Sets, and Evidence Theory" (PDF). Creighton University. Retrieved 16 July 2016.

11. Gerla,, G. (2016). "Comments on some theories of fuzzy computation". International Journal of general Systems,. 45 *(4): 372–392.*

12. Novák, V (2005). "Are fuzzy sets a reasonable tool for modeling vague phenomena?". *Fuzzy Sets and Systems.* 156: 341–348. doi:10.1016/j.fss.2005.05.029.

13. Valiant, Leslie, (2013) Probably Approximately Correct: Nature"s Algorithms for Learning and Prospering in a Complex World New York: Basic Books. ISBN 978-0465032716

14. Cejas, Jesús, (2011) Compensatory Fuzzy Logic. La Habana: Revista de Ingeniería Industrial. ISSN 1815-5936

CHAPTER 2

Fuzzy Concept

Chapter Contents

- Introduction
- Origin And Etymology
- Applications
- Fuzzy Concept Lattices And Big Data
- Definitional Controversies
- Sociology And Journalism
- Uncertainty
- Language
- Psychology
- Everyday Occurrence
- Economy Of Distinctions
- Analysis

Formal concept analysis (FCA) deals with a particular kind of analysis of data which, in the basic setting, has the form of a table with rows corresponding to objects, columns corresponding to attributes, and table entries containing 1's and 0's depending on whether an object has or does not have an attribute (we assume basic familiarity with FCA and refer to for information). The basic setting is well-suited for attributes which are crisp, i.e. each object of the domain of applicability of the attribute either has (1) or does not have (0) the attribute. Many attributes are fuzzy rather than crisp. That is to say, it is a matter of degree to which an object has a (fuzzy) attribute. For instance, when asking whether a man with a height of 182 cm is tall, one probably gets an answer like "not completely tall but almost tall" or "to a high degree tall", etc. A natural idea, developed in fuzzy logic], is to assign to an object a truth degree to which the object has a (fuzzy) attribute. Degrees are taken from an appropriate scale L of truth degrees. A favorite choice of L is the real unit interval [0, 1] or some subset of [0, 1]. Then, we can say that a man with a height of 182 cm is tall to a degree, say, 0.8. Doing so, the entries of a table describing objects and attributes become degrees from L instead of values from {0, 1} as is the case of the basic setting of FCA.

INTRODUCTION

A **fuzzy concept** is a concept of which the boundaries of application can vary considerably according to context or conditions, instead of being fixed once and for all. This means the concept is *vague* in some way, lacking a fixed, precise meaning, without however being unclear or meaningless altogether. It has a definite meaning, which can become more precise only through further elaboration and specification, including a closer definition of the context in which the concept is used.

A fuzzy concept is understood by scientists as a concept which is "to an extent applicable" in a situation, and it therefore implies gradations of meaning. The best known example of a fuzzy concept around the world is an amber traffic light, and indeed fuzzy concepts are nowadays widely used in traffic control systems.

The Nordic myth of Loki"s wager suggests that concepts which lack a precise meaning or precise boundaries of application cannot be usefully discussed at all. However, the idea of "fuzzy concepts" proposes that "somewhat vague terms" can be operated with, since we can explicate and define the variability of their application, by assigning numbers to it.

ORIGIN AND ETYMOLOGY

The intellectual origins of the idea of fuzzy concepts have been traced to a diversity of famous and less well known thinkers including Plato, Georg Wilhelm Friedrich Hegel, Karl Marx, Friedrich Engels, Friedrich Nietzsche, Jan Łukasiewicz, Alfred Tarski, Stanisław Jaśkowski and Donald Knuth. This suggests that the idea of fuzzy concepts has, in one form or another, a very long history in human thought.

However, usually the Iranian born, American computer scientist Lotfi A. Zadeh is credited with inventing the specific idea of a "fuzzy concept" in his seminal 1965 paper on fuzzy sets, because he gave a formal mathematical presentation of the phenomenon which was widely accepted by scholars. It was also Zadeh who played a decisive role in developing the field of fuzzy logic, fuzzy sets and fuzzy systems, with a large number of scholarly papers.

In fact, the German scholar Dieter Klaua also published a German-language paper on fuzzy sets in 1965, but he used a different terminology (he referred to "many-valued sets"). An earlier attempt to create a theory of sets where set membership is a matter of degree was made by Abraham Kaplan and Hermann Schott in 1951. They intended to apply the idea to empirical research. Kaplan and Schott measured the degree of membership of empirical classes using real numbers between 0 and 1, and they defined corresponding notions of intersection, union, complementation and subset. However, at the time, their idea "fell on stony ground".

Radim Belohlavek explains:

"There exists strong evidence, established in the 1970s in the psychology of concepts... that human concepts have a graded structure in that whether or not a concept applies to a given object is a matter of degree, rather than a yes-or-no question, and that people are capable of working with the degrees in a consistent way. This finding is intuitively quite appealing, because people say "this product is more or less good" or "to a certain degree, he is a good athlete", implying the graded structure of concepts. In his classic paper, Zadeh called the concepts with a graded structure *fuzzy concepts* and argued that these concepts are a rule rather than an exception when it comes to how people communicate knowledge. Moreover, he argued that to model such concepts mathematically is important for the tasks of control, decision making, pattern recognition, and the like. Zadeh proposed the notion of a *fuzzy set* that gave birth to the field of *fuzzy logic*..."

Hence, a concept is regarded as "fuzzy" in a logical sense if:

- defining characteristics of the concept apply to it "to a certain degree or extent" (or, more unusually, "with a certain magnitude of likelihood")
- or, the boundaries of applicability (the truth-value) of a concept can vary according to different conditions.
- or, the fuzzy concept itself straightforwardly consists of a fuzzy set, or a combination of such sets.

The fact that a concept is fuzzy does not prevent its use in logical reasoning, it merely affects the type of reasoning which can be applied (see fuzzy logic). If the concept has gradations of meaningful significance, it is necessary to specify and formalize what those gradations are. Not all fuzzy concepts have the same logical structure, but they can often be formally described or reconstructed using fuzzy logic.

Zadeh's seminal 1965 paper is acknowledged to be one of the most-cited scholarly articles in the 20th century.

APPLICATIONS

In philosophical logic, fuzzy concepts are often regarded as concepts which in their application, or formally speaking, are neither completely true nor completely false, or which are partly true and partly false; they are ideas which require further elaboration, specification or qualification to understand their applicability (the conditions under which they truly make sense).

In mathematics and statistics, a fuzzy variable (such as "the temperature", "hot" or "cold") is a value which could lie in a probable range defined by quantitative limits or parameters, and which can be usefully described with imprecise categories (such as "high", "medium" or "low") using some kind of qualitative scale.

In mathematics and computer science, the gradations of applicable meaning of a fuzzy concept are described in terms of *quantitative* relationships defined by logical operators. Such an approach is sometimes called "degree-theoretic semantics" by logicians and philosophers, but the more usual term is fuzzy logic or many-valued logic. The novelty of fuzzy logic is, that it "breaks with the traditional principle that formalisation should correct and avoid, but not compromise with, vagueness".

The basic idea of fuzzy logic is, that a real number is assigned to each statement written in a language, within a range from 0 to 1, where 1 means that the statement is completely true, and 0 means that the statement is completely false, while values less than 1 but greater than 0 represent that the statements are "partly true", to a given, quantifiable extent. Susan Haack comments:

"Whereas in classical set theory an object either is or is not a member of a given set, in fuzzy set theory membership is a matter of degree; the degree of membership of an object in a fuzzy set is represented by some real number between 0 and 1, with 0 denoting *no* membership and 1 *full* membership."

"Truth" in this mathematical context usually means simply that "something is the case", or that "something is applicable". This makes it possible to analyze a distribution of statements for their truth-content, identify data patterns, make inferences and predictions, and model how processes operate. Fuzzy logic in principle allows us to give a definite, precise answer to the question: "to what extent is something the case?", or "to what extent is something applicable?". Via a series of switches, this kind of reasoning can be built into electronic devices. That was already happening before fuzzy logic was invented, but using fuzzy logic in modelling has become an important aid in design, which creates many new technical possibilities.

Fuzzy reasoning (i.e., reasoning with graded concepts) turns out to have many practical uses. It is nowadays widely used in the programming of vehicle and transport electronics, household appliances, video games, language filters, robotics, all kinds of control systems, and various kinds of electronic equipment used for pattern recognition, surveying and monitoring (such as radars). Fuzzy reasoning is also used in artificial intelligence, virtual intelligence and soft computing research. "Fuzzy risk scores" are used by project managers and portfolio managers to express risk assessments. Fuzzy logic has even been applied to the problem of predicting cement strength. It looks like fuzzy logic will eventually be applied in almost every aspect of life, even if people are not aware of it, and in that sense fuzzy logic is an astonishingly successful invention.

A lot of research on fuzzy logic was done by Japanese researchers inventing new machinery, electronic equipment and appliances (see also Fuzzy control system). The North American Fuzzy Information Processing Society (NAFIPS) was founded in 1981. There also exists an International Fuzzy Systems Association (IFSA). In Europe, there is a European Society for Fuzzy Logic and Technology (EUSFLAT).

Lotfi Zadeh estimates there are more than 50,000 fuzzy logic-related, patented inventions. He lists 28 journals dealing with fuzzy reasoning, and 21 journal titles on soft computing. There are now close to 100,000 publications with the word "fuzzy" in their titles, or maybe even 300,000.

FUZZY CONCEPT LATTICES AND BIG DATA

According to the computer scientist Andrei Popescu at Middlesex University London, a concept can be operationally defined to consist of (1) an *intent*, which is a description or specification stated in a language, and (2) an *extent*, which is the collection of all the objects to which the description refers. Additionally the concept can be defined only with reference to a *context*, which is stated by (i) the universe of all possible objects, (ii) the universe of all possible attributes of objects, and (iii) the formal definition of the relation whereby an object possesses an attribute. Once the context is defined, we can then specify relationships between sets of objects and sets of attributes which they do, or do not share.

However, whether an object belongs to a concept, and whether an object does, or does not have an attribute, can often be a matter of degree. Thus, for example, "many attributes are fuzzy rather than crisp". To overcome this issue, a numerical value is assigned to each object or attribute along a scale, and the results are placed in a table which links each assigned value within the given range to a numerical value denoting a given degree of applicability.

This is the basic idea of a "fuzzy concept lattice", which can also be graphed; different fuzzy concept lattices can be connected to each other as well. Fuzzy concept lattices are a useful programming tool for the exploratory analysis of big data, for example in cases where sets of linked behavioural responses are broadly similar, but can nevertheless vary in important ways, within certain limits. It can help to find out what the structure and dimensions are, of a behaviour that occurs with an important but limited amount of variation within a large population.

Fuzzy lattices can be applied, for instance, in the psephological analysis of big data about voter behaviour, where researchers want to explore the characteristics and associations involved in "somewhat vague" opinions; patterns of variation in voter attitudes; and variability in voter behaviour (or personal characteristics) within a set of parameters. The basic programming techniques for this kind of fuzzy concept mapping and deep learning are by now well-established and big data analytics had a strong influence on the US elections of 2016. A US study concluded in 2015 that for 20% of undecided voters, Google"s secret search algorithm had the power to change the way they voted.

Very large quantities of data can now be explored using computers with fuzzy logic systems and open-source architectures such as Apache Hadoop, Apache Spark, and MongoDB. One author claimed in 2016 that it is now possible to obtain and analyze "400 data points" for each voter in a population, using Oracle systems. However, NBC News reported that the British firm Cambridge Analytica which profiled voters for Donald Trump has not 400, but 4,000 data points for each of 230 million US adults. Cambridge Analytica"s own website claims that "up to 5,000 data points" are collected for each of 220 million Americans. Harvard University Professor Latanya Sweeney calculated, that if a company knows just your date of birth, your ZIP code and sex, the company can identify you by name 87% of the time - simply by using linked data from various sources. With 4,000-5,000 data points instead of three, a very comprehensive personal profile becomes possible for almost every voter, and many behavioural patterns can be inferred by linking together different data sets. It also becomes possible to identify and measure gradations in personal characteristics which, in aggregate, have very large effects.

Some researchers argue that this kind of big data analysis has severe limitations, and that the analytical results can only be regarded as indicative, and not as definitive. This was confirmed by Kellyanne Conway, Donald Trump's campaign manager, who emphasized the importance of human judgement and common sense in drawing conclusions from fuzzy data. Conway candidly admitted that much of her own research would "never see the light of day", because it was client confidential. Another Trump adviser criticized Conway, claiming that she "produces an analysis that buries every terrible number and highlights every positive number". A traditional objection to big data is, that it cannot cope with rapid change, but the technology now exists for corporations like Amazon, Google and Microsoft to pump cloud-based data streams from app-users straight into big data analytics programmes, in real time.

DEFINITIONAL CONTROVERSIES

Some scientists claimed that in reality fuzzy concepts do not exist. For example, Rudolf E. Kálmán stated in 1972 that "there is no such thing as a fuzzy concept... We do talk about fuzzy things but they are not scientific concepts". The suggestion is that a concept, to qualify as a concept, must be clear and precise. A vague notion would be at best a prologue to formulating a concept. However, there is no general agreement how the notion of a "concept", or a scientific concept in particular, should be defined, and of course scientists also quite often do use imprecise analogies in their models to help understanding an issue.

Susan Haack once claimed that a many-valued logic requires neither intermediate terms between true and false, nor a rejection of bivalence. Her suggestion was, that the intermediate terms (i.e. the gradations of truth) can always be restated as conditional if-then statements, and by implication, that fuzzy logic is fully reducible to binary true-or-false logic. This interpretation is disputed, but even if it was correct, the ability to assign a numerical value to the applicability of a statement is often enormously more efficient than a long sequence of if-then statements that would have the same meaning. That point is obviously of great importance to computer programmers seeking to code a process or operation according to logical rules.

- Not all scholars would agree that a concept is equal to, or reducible to, a mathematical set.
- Qualities may not be fully reducible to quantities - if there are no qualities, it becomes impossible to say what the numbers are numbers of, or what they refer to, except that they refer to other numbers.
- In creating a formalization or formal specification of a concept, for example for the purpose of measurements, administrative procedure or programming, part of the meaning of the concept may be changed or lost.

Programmers, statisticians or logicians are concerned in their work with the main operational or technical significance of a concept which is specifiable in objective, quantifiable terms. They are not primarily concerned with all kinds of interpretive frameworks associated with the concept, or with those aspects of the concept which seem to have no particular functional purpose - however entertaining they might be. However, some of the qualitative characteristics of the concept may not be quantifiable or measurable at all, at least not directly. The temptation exists to ignore them, or try to infer them from data results. A concept may also be created as an ideal type to understand something imaginatively, without any strong claim that it is a "true and complete description" or a "true and complete reflection" of whatever is being conceptualized.

Philosophers regard fuzziness as a particular kind of vagueness, and consider that "no specific assignment of semantic values to vague predicates, not even a fuzzy one, can fully satisfy our conception of what the extensions of vague predicates are like". However, Lotfi

Zadeh claims that "vagueness connotes insufficient specificity, whereas fuzziness connotes unsharpness of class boundaries". Thus, he argues, a sentence like "I will be back in a few minutes" is fuzzy *but not* vague, whereas a sentence such as "I will be back sometime", is fuzzy *and* vague. His suggestion is that fuzziness and vagueness are logically quite different qualities, rather than fuzziness being a type or subcategory of vagueness. Zadeh claims that "inappropriate use of the term ‹vague" is still a common practice in the literature of philosophy".

The definitional disputes remain unresolved, mainly because, as anthropologists and psychologists have documented, different languages (or symbol systems) that have been created by people to signal meanings suggest different ontologies. Put simply: it is not merely that describing "what is there" involves symbolic representations of some kind. How distinctions are drawn, influences perceptions of "what is there", and vice versa, perceptions of "what is there" influence how distinctions are drawn. For example, cosmologist Max Tegmark argues the universe consists of math: "If you accept the idea that both space itself, and all the stuff in space, have no properties at all except mathematical properties," then the idea that everything is mathematical "starts to sound a little bit less insane."

SOCIOLOGY AND JOURNALISM

The idea of fuzzy concepts has also been applied in the philosophical, sociological and linguistic analysis of human behaviour. In a 1973 paper, George Lakoff for example analyzed hedges in the interpretation of the meaning of categories. Charles Ragin and others have applied the idea to sociological analysis.

In a more general sociological or journalistic sense, a "fuzzy concept" has come to mean a concept which is meaningful but inexact, implying that it does not exhaustively or completely define the meaning of the phenomenon to which it refers - often because it is too abstract. In this context, it is said that fuzzy concepts "lack clarity and are difficult to test or operationalize".To specify the relevant meaning more precisely, additional distinctions, conditions and/or qualifiers would be required. Thus, for example, in a handbook of sociology we find a statement such as "The theory of interaction rituals contains some gaps that need to be filled and some fuzzy concepts that need to be differentiated." The idea is that if finer distinctions are introduced, then the fuzziness or vagueness would be eliminated.

The main reason why the term is now often used in describing human behaviour, is that human interaction has many characteristics which are difficult to quantify and measure precisely, although we know that they have magnitudes, among other things because they are interactive and reflexive (the observers and the observed mutually influence the meaning of events). Those human characteristics can be usefully expressed only in an *approximate* way (see reflexivity (social theory)).

Newspaper stories frequently contain fuzzy concepts, which are readily understood and used, even although they are far from exact. Thus, many of the meanings which people ordinarily use to negotiate their way through life in reality turn out to be "fuzzy concepts". While people often do need to be exact about some things (e.g. money or time), many areas of their lives involve expressions which are far from exact.

Sometimes the term is also used in a pejorative sense. For example, a New York Times journalist wrote that Prince Sihanouk "seems unable to differentiate between friends and enemies, a disturbing trait since it suggests that he stands for nothing beyond the fuzzy concept of peace and prosperity in Cambodia".

The use of fuzzy logic in the social sciences and humanities has remained limited. Lotfi Zadeh said in a 1994 interview that:

"I expected people in the social sciences - economics, psychology, philosophy, linguistics, politics, sociology, religion and numerous other areas to pick up on it. It"s been somewhat of a mystery to me why even to this day, so few social scientists have discovered how useful it could be."

Two decades later, fuzzy concepts and fuzzy logic are being widely applied in big data analysis of social, commercial and psychological phenomena. Jaakko Hintikka once claimed that "the logic of natural language we are in effect already using can serve as a "fuzzy logic" better than its trade name variant without any additional assumptions or constructions." That might help to explain why fuzzy logic is not much used to formalize concepts in the "soft" social sciences. However Lotfi Zadeh rejected such an interpretation, on the ground that in many human endeavours as well as technologies it is highly important to define more exactly "to what extent" something is applicable or true, when it is known that its applicability can vary to some important extent. Reasoning which accepts and uses fuzzy concepts can be shown to be perfectly valid with the aid of fuzzy logic, because the degrees of applicability of a concept can be more precisely and efficiently defined with the aid of numerical notation.

Another possible explanation for the lack of use of fuzzy logic by social scientists is simply that, beyond basic statistical analysis (using programs such as SPSS and Excel) the mathematical knowledge of social scientists is often rather limited; they may not know how to formalize a fuzzy concept using the conventions of fuzzy logic. But Hintikka may be correct, in the sense that it can be much more efficient to use natural language to denote a complex idea, than to formalize it. The quest for formalization might introduce much more complexity, which is not wanted, and which detracts from communicating the relevant issue.

UNCERTAINTY

Fuzzy concepts can generate uncertainty because they are imprecise (especially if they refer to a process in motion, or a process of transformation where something is "in the process of turning into something else"). In that case, they do not provide a clear orientation for action or decision-making ("what does X really mean or imply?"); reducing fuzziness, perhaps by applying fuzzy logic, would generate more certainty.

However, this is not necessarily always so. A concept, even although it is not fuzzy at all, and even though it is very exact, could equally well fail to capture the meaning of something adequately. That is, a concept can be very precise and exact, but not - or insufficiently - *applicable* or *relevant* in the situation to which it refers. In this sense, a definition can be "very precise", but "miss the point" altogether.

A fuzzy concept may indeed provide *more* security, because it provides a meaning for something when an exact concept is unavailable - which is better than not being able to denote it at all. A concept such as God, although not easily definable, for instance can provide security to the believer.

LANGUAGE

Ordinary language, which uses symbolic conventions and associations which are often not logical, inherently contains many fuzzy concepts - "knowing what you mean" in this case depends on knowing the context or being familiar with the way in which a term is normally used, or what it is associated with. This can be easily verified for instance by consulting a dictionary, a thesaurus or an encyclopedia which show the multiple meanings of words, or by observing the behaviours involved in ordinary relationships which rely on mutually understood meanings. Bertrand Russell regarded language as intrinsically vague.

To communicate, receive or convey a message, an individual somehow has to bridge his own intended meaning and the meanings which are understood by others, i.e., the message has to be conveyed in a way that it will be socially understood, preferably in the intended manner. Thus, people might state: "you have to say it in a way that I understand".

This may be done instinctively, habitually or unconsciously, but it usually involves a choice of terms, assumptions or symbols whose meanings may often not be completely fixed, but which depend among other things on how the receiver of the message responds to it, or the context. In this sense, meaning is often "negotiated" or "interactive" (or, more cynically, manipulated). This gives rise to many fuzzy concepts.

But even using ordinary set theory and binary logic to reason something out, logicians have discovered that it is possible to generate statements which are logically speaking not completely true or imply a paradox, even although in other respects they conform to logical rules.

PSYCHOLOGY

The formation of fuzzy concepts is partly due to the fact that the human brain does not operate like a computer (see also Chinese room).

- While ordinary computers use strict binary logic gates, the brain does not; i.e., it is capable of making all kinds of neural associations according to all kinds of ordering principles (or fairly chaotically) in associative patterns which are not logical but nevertheless meaningful. For example, a work of art can be meaningful without being logical. A pattern can be regular and non-arbitrary, hence meaningful, without it being possible to describe it completely or exhaustively in formal-logical terms.
- Something can be meaningful although we cannot name it, or we might only be able to name it and nothing else.
- The human brain can also interpret the same phenomenon in several different but interacting frames of reference, at the same time, or in quick succession, without there necessarily being an explicit logical connection between the frames.

In part, fuzzy concepts arise also because learning or the growth of understanding involves a transition from a vague awareness, which cannot orient behaviour greatly, to clearer insight, which can orient behaviour.

Some logicians argue that fuzzy concepts are a necessary consequence of the reality that any kind of distinction we might like to draw has *limits of application*. At a certain level of generality, a distinction works fine. But if we pursued its application in a very exact and rigorous manner, or overextend its application, it appears that the distinction simply does not apply in some areas or contexts, or that we cannot fully specify how it should be drawn. An analogy might be that zooming a telescope, camera, or microscope in and out reveals that a pattern which is sharply focused at a certain distance disappears at another distance (or becomes blurry).

Faced with any large, complex and continually changing phenomenon, any short statement made about that phenomenon is likely to be "fuzzy", i.e., it is meaningful, but - strictly speaking - incorrect and imprecise. It will not really do full justice to the reality of what is happening with the phenomenon. A correct, precise statement would require a lot of elaborations and qualifiers. Nevertheless, the "fuzzy" description turns out to be a useful shorthand that saves a lot of time in communicating what is going on ("you know what I mean").

In psychophysics it has been discovered that the perceptual distinctions we draw in the mind are often more sharply defined than they are in the real world. Thus, the brain actually tends to "sharpen up" our perceptions of differences in the external world. Between black and white, we are able to detect only a limited number of shades of gray, or colour gradations. If there are more gradations and transitions in reality, than our conceptual or perceptual distinctions can capture, then it could be argued that how those distinctions will actually apply, must *necessarily* become vaguer at some point. If, for example, one wants to count and quantify distinct objects using numbers, one needs to be able to distinguish between those separate objects, but if this is difficult or impossible, then, although this may not invalidate a quantitative procedure as such, quantification is not really possible in practice; at best, we may be able to assume or infer indirectly a certain distribution of quantities that must be there.

Finally, in interacting with the external world, the human mind may often encounter new, or partly *new phenomena or relationships* which cannot (yet) be sharply defined given the background knowledge available, and by known distinctions, associations or generalizations.

"Crisis management plans cannot be put ‹on the fly" after the crisis occurs. At the outset, information is often vague, even contradictory. Events move so quickly that decision makers experience a sense of loss of control. Often denial sets in, and managers unintentionally cut off information flow about the situation" - L. Paul Bremer, "Corporate governance and crisis management", in: *Directors & Boards*, Winter 2002

It also can be argued that fuzzy concepts are generated by a certain sort of lifestyle or way of working which evades definite distinctions, makes them impossible or inoperable, or which is in some way chaotic. To obtain concepts which are not fuzzy, it must be possible to test out their application in some way. But in the absence of any relevant clear distinctions, or when everything is "in a state of flux" or in transition, it may not be possible to do so, so that the amount of fuzziness increases.

EVERYDAY OCCURRENCE

Fuzzy concepts often play a role in the creative process of forming new concepts to understand something. In the most primitive sense, this can be observed in infants who, through practical experience, learn to identify, distinguish and generalise the correct application of a concept, and relate it to other concepts.

However, fuzzy concepts may also occur in scientific, journalistic, programming and philosophical activity, when a thinker is in the process of clarifying and defining a newly emerging concept which is based on distinctions which, for one reason or another, cannot (yet) be more exactly specified or validated. Fuzzy concepts are often used to denote complex phenomena, or to describe something which is developing and changing, which might involve shedding some old meanings and acquiring new ones.

- In politics, it can be highly important and problematic how exactly a conceptual distinction is drawn, or indeed whether a distinction is drawn at all; distinctions used in administration may be deliberately sharpened, or kept fuzzy, due to some political motive or power relationship. Politicians may be deliberately vague about some things, and very clear and explicit about others; if there is information that proves their case, they become very precise, but if the information doesn"t prove their case, they become vague or saying nothing. The "fuzzy area" can also refer simply to a *residual* number of cases which cannot be allocated to a known and identifiable group, class or set if strict criteria are used.
- In administration and accounting, fuzziness problems of interpretation and boundary problems can arise, because it is not clear to what category exactly a case, item, transaction or piece of information belongs. In principle, each case, event or item must be allocated to the correct category in a procedure, but it may be, that it is difficult to make the appropriate or relevant distinctions.
- In translation work, fuzzy concepts are analyzed for the purpose of good translation. A concept in one language may not have quite the same meaning or significance in another language, or it may not be feasible to translate it literally, or at all. Some languages have concepts which do not exist in another language, raising the problem of how one would most easily render their meaning. In computer-assisted translation, a technique called fuzzy matching is used to find the most likely translation of a piece of text, using previous translated texts as a basis.
- In information services fuzzy concepts are frequently encountered because a customer or client asks a question about something which could be interpreted in many different ways, or, a document is transmitted of a type or meaning which cannot be easily allocated to a known type or category, or to a known procedure. It might take considerable inquiry to "place" the information, or establish in what framework it should be understood.

- In the legal system, it is essential that rules are interpreted and applied in a standard way, so that the same cases and the same circumstances are treated equally. Otherwise one would be accused of arbitrariness, which would not serve the interests of justice. Consequently, lawmakers aim to devise definitions and categories which are sufficiently precise, so that they are not open to different interpretations. For this purpose, it is critically important to remove fuzziness, and differences of interpretation are typically resolved through a court ruling based on evidence. Alternatively, some other procedure is devised which permits the correct distinction to be discovered and made.
- In medical diagnosis, the assessment of what the symptoms of a patient are often cannot be very exactly specified, since there are many possible qualitative and quantitative gradations in severity, incidence or frequency that could occur. Different symptoms may also overlap to some extent. These gradations can be difficult to measure, and so the medical professionals use approximate "fuzzy" categories in their judgement of a medical condition or a patient"s state of health. Although it may not be exact, the diagnosis is often useful enough for treatment purposes.
- In statistical research, it is an aim to measure the magnitudes of phenomena. For this purpose, phenomena have to be grouped and categorized so that distinct and discrete counting units can be defined. It must be possible to allocate all observations to mutually exclusive categories so that they are properly quantifiable. Survey observations do not spontaneously transform themselves into countable data; they have to be identified, categorized and classified in such a way, that identical observations can be grouped together, and that observations are not counted twice or more. Again, for this purpose it is a requirement that the concepts used are exactly defined, and not fuzzy. There could be a margin of error, but the amount of error must be kept within tolerable limits, and preferably its magnitude should be known.
- In hypnotherapy, fuzzy language is deliberately used for the purpose of trance induction. Hypnotic suggestions are often couched in a somewhat vague, general or ambiguous language requiring interpretation by the subject. The intention is to distract and shift the conscious awareness of the subject away from external reality to his own internal state. In response to the somewhat confusing signals he gets, the awareness of the subject spontaneously tends to withdraw inward, in search of understanding or escape.
- In biology, protein complexes with multiple structural forms are called fuzzy complexes. The different conformations can result in different, even opposite functions. The conformational ensemble is modulated by the environmental conditions. Post-translational modifications or alternative splicing can also impact the ensemble and thereby affinity or specificity of interactions.
- In theology an attempt is made to define more precisely the meaning of spiritual concepts, which refer to how human beings construct the meaning of human existence, and, often, the relationship people have with a supernatural world. Many spiritual concepts and beliefs are fuzzy, to the

extent that, although abstract, they often have a highly personalized meaning, or involve personal interpretation of a type that is not easy to define in a cut-and-dried way. A similar situation occurs in psychotherapy. The Dutch theologian Kees de Groot has explored the imprecise notion that psychotherapy is like an "implicit religion", defined as a "fuzzy concept" (it all depends on what one means by "psychotherapy" and "religion").

- In meteorology, where changes and effects of complex interactions in the atmosphere are studied, the weather reports often use fuzzy expressions indicating a broad trend, likelihood or level. The main reason is that the forecast can rarely be totally exact for any given location.
- In phenomenology which studies the structure of subjective experience, an important insight is that how someone experiences something can be influenced *both* by the influence of the thing being experienced itself, but *also* by how the person responds to it. Thus, the actual experience the person has, is shaped by an "interactive object-subject relationship". To describe this experience, fuzzy categories are often necessary, since it is often impossible to predict or describe with great exactitude what the interaction will be, and how it is experienced.

It could be argued that many concepts used fairly universally in daily life (e.g. "love" or "God" or "health" or "social") are *inherently or intrinsically* fuzzy concepts, to the extent that their meaning can never be completely and exactly specified with logical operators or objective terms, and can have multiple interpretations, which are in part exclusively subjective. Yet despite this limitation, such concepts are not meaningless. People keep using the concepts, even if they are difficult to define precisely.

It may also be possible to specify one personal meaning for the concept, without however placing restrictions on a different use of the concept in other contexts (as when, for example, one says "this is what I mean by X" in contrast to other possible meanings). In ordinary speech, concepts may sometimes also be uttered purely randomly; for example a child may repeat the same idea in completely unrelated contexts, or an expletive term may be uttered arbitrarily. A feeling or sense is conveyed, without it being fully clear what it is about.

Fuzzy concepts can be used deliberately to create ambiguity and vagueness, as an evasive tactic, or to bridge what would otherwise be immediately recognized as a contradiction of terms. They might be used to indicate that there is definitely a connection between two things, without giving a complete specification of what the connection is, for some or other reason. This could be due to a failure or refusal to be more precise. But it could also could be a prologue to a more exact formulation of a concept, or a better understanding.

ECONOMY OF DISTINCTIONS

Fuzzy concepts can be used as a practical method to describe something of which a complete description would be an unmanageably large undertaking, or very time-consuming; thus, a simplified indication of what is at issue is regarded as sufficient, although it is not exact. There is also such a thing as an "economy of distinctions", meaning that it is not helpful or efficient to use more detailed definitions than are really necessary for a given purpose. The provision of "too many details" could be disorienting and confusing, instead of being enlightening, while a fuzzy term might be sufficient to provide an orientation. The reason for using fuzzy concepts can therefore be purely pragmatic, if it is not feasible or desirable (for practical purposes) to provide "all the details" about the meaning of a shared symbol or sign. Thus people might say "I realize this is not exact, but you know what I mean" - they assume practically that stating all the details is not required for the purpose of the communication.

Lotfi Zadeh has picked up this point, and draws attention to a "major misunderstanding" about applying fuzzy logic. It is true that the basic aim of fuzzy logic is to make what is imprecise more precise. Yet in many cases, fuzzy logic is used paradoxically to "imprecisiate what is precise", meaning that there is a deliberate tolerance for imprecision for the sake of simplicity of procedure and economy of expression. In such uses, there is a tolerance for imprecision, because making ideas more precise would be unnecessary and costly, while "imprecisiation reduces cost and enhances tractability" (tractability means "being easy to manage or operationalize"). Zadeh calls this approach the "Fuzzy Logic Gambit" (a gambit means giving up something now, to achieve a better position later). In the Fuzzy Logic Gambit, "what is sacrificed is precision in [quantitative] value, but not precision in meaning", and more concretely, "imprecisiation in value is followed by precisiation in meaning". He cites as example Takeshi Yamakawa"s programming for an inverted pendulum, where differential equations are replaced by fuzzy if-then rules in which words are used in place of numbers.

ANALYSIS

In mathematical logic, computer programming, philosophy and linguistics fuzzy concepts can be analyzed and defined more accurately or comprehensively, by describing or modelling the concepts using the terms of fuzzy logic. More generally, clarification techniques can be used such as:

- concretizing the concept by finding specific examples, illustrations, details or cases to which it applies.
- specifying a range of *conditions* to which the concept applies (for example, in computer programming of a procedure).
- *classifying* or *categorizing* all or most cases or uses to which the concept applies (taxonomy).
- *probing the assumptions* on which a concept is based, or which are associated with its use (Critical thought).
- identifying *operational rules* for the use of the concept, which cover all or most cases.
- allocating different applications of the concept to different but related *sets* (e.g. using Boolean logic).
- examining how *probable* it is that the concept applies, statistically or intuitively (Probability theory).

- examining the *distribution* or distributional frequency of (possibly different) uses of the concept (statistics).
- some other kind of *measure or scale* of the degree to which the concept applies.
- specifying a series of logical operators (an inferential system or algorithm) which captures all or most cases to which the concept applies.
- *mapping or graphing* the applications of the concept using some basic parameters.
- applying a *meta-language* which includes fuzzy concepts in a more inclusive categorical system which is not fuzzy.
- *reducing or restating* fuzzy concepts in terms which are simpler or similar, and which are not fuzzy or less fuzzy.
- *relating* the fuzzy concept to other concepts which are not fuzzy or less fuzzy, or simply by *replacing* the fuzzy concept altogether with another, alternative concept which is not fuzzy yet "works exactly the same way".

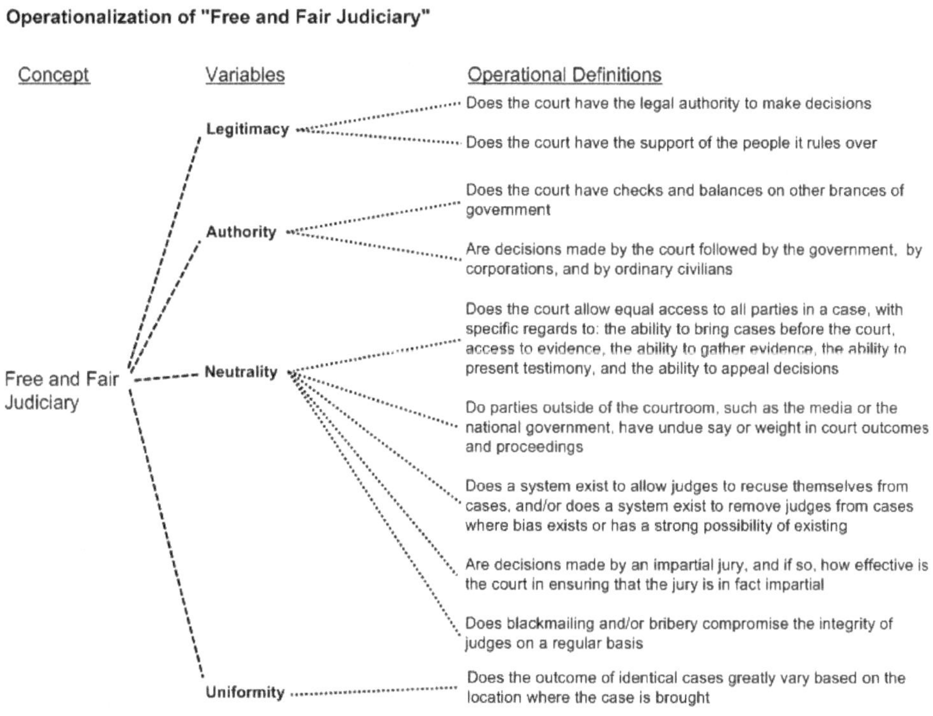

An operationalization diagram, one method of clarifing fuzzy concepts.

In this way, we can obtain a more exact understanding of the use of a fuzzy concept, and possibly decrease the amount of fuzziness. It may not be possible to specify all the possible meanings or applications of a concept completely and exhaustively, but if it is possible to capture the majority of them, statistically or otherwise, this may be useful enough for practical purposes.

A process of defuzzification is said to occur, when fuzzy concepts can be logically described in terms of (the relationships between) fuzzy sets, which makes it possible to define variations in the meaning or applicability of concepts as quantities. Effectively, qualitative differences may then be described more precisely as quantitative variations or quantitative variability (assigning a numerical value then denotes the magnitude of variation).

The difficulty that can occur in judging the fuzziness of a concept can be illustrated with the question *"Is this one of those?"*. If it is not possible to clearly answer this question, that could be because "this" (the object) is itself fuzzy and evades definition, or because "one of those" (the concept of the object) is fuzzy and inadequately defined.

Thus, the source of fuzziness may be in the nature of the reality being dealt with, the concepts used to interpret it, or the way in which the two are being related by a person. It may be that the personal meanings which people attach to something are quite clear to the persons themselves, but that it is not possible to communicate those meanings to others except as fuzzy concepts.

REFERENCES

1. Susan Haack, *Deviant logic, fuzzy logic: beyond the formalism*. Chicago: University of Chicago Press, 1996.
2. Richard Dietz & Sebastiano Moruzzi (eds.), *Cuts and clouds. Vagueness, Its Nature, and Its Logic*. Oxford University Press, 2009.
3. Sandeep Mehan & Vandana Sharma, "Development of traffic light control system based on fuzzy logic". *ACAI '11 Proceedings of the International Conference on Advances in Computing and Artificial Intelligence 2011*, pp. 162-165.
4. Susan Haack notes that Stanisław Jaśkowski provided axiomatizations of many-valued logics in: Jaśkowski, "On the rules of supposition in formal logic. *Studia Logica* No. 1, 1934. See Susan Haack, *Philosophy of Logics*. Cambridge University Press, 1978, p. 205.
5. Priyanka Kaushal, Neeraj Mohan and Parvinder S. Sandhu, "Relevancy of Fuzzy Concept in Mathematics". *International Journal of Innovation, Management and Technology*, Vol. 1, No. 3, August 2010.
6. Lotfi A. Zadeh, "Fuzzy sets". In: *Information and Control*, Vol. 8, June 1965, pp. 338–353.
7. Lotfi A Zadeh with George J. Klir and Bo Yua, *Fuzzy Sets, Fuzzy Logic, and Fuzzy Systems: Selected Papers*. Singapore and River Edge (N.J.): World Scientific Publishing Company, 1996.
8. Dieter Klaua. "Über einen Ansatz zur mehrwertigen Mengenlehre". *Monatsberichte der Deutschen Akademie der Wissenschaften* (Berlin), Vol. 7: pp. 859-867, 1965.
9. Siegfried Gottwald, "Shaping the logic of fuzzy set theory". In: Cintula, Petr et al. (eds.), *Witnessed years. Essays in honour of Petr Hájek*. London: College Publications, 2009, pp. 193-208.

10. Abraham Kaplan and Hermann F. Schott, "A calculus for empirical classes", Methodos Vol. 3, 1951, pp. 165–188.

11. Timothy Williamson, *Vagueness*. London: Routledge, 1996, p. 120.

12. Radim Belohlavek, "What is a fuzzy concept lattice? II", in: Sergei O. Kuznetsov et al. (eds.), *Rough sets, fuzzy sets, data mining and granular computing*. Berlin: Springer Verlag, 2011, pp. 19-20.

13. *IFSA Newsletter* (International Fuzzy Systems Association), Vol. 10, No. 1, March 2013

14. Roy T. Cook, *A dictionary of philosophical logic*. Edinburgh University Press, 2009, p. 84.

15. Susan Haack, *Philosophy of Logics*. Cambridge University Press, 1978, p. xii.

16. Susan Haack, *Philosophy of Logics*. Cambridge University Press, 1978, p. 165.

17. Kazuo Tanaka, *An Introduction to Fuzzy Logic for Practical Applications*. Springer, 1996; Constantin Zopounidis, Panos M. Pardalos & George Baourakis, *Fuzzy Sets in Management, Economics and Marketing*. Singapore; World Scientific Publishing Co. 2001.

18. *Stosberg, Mark (16 December 1996). "The Role of Fuzziness in Artifical [sic] Intelligence". Minds and Machines. Retrieved 19 April 2013.*

19. Irem Dikmen, M. Talat Birgonal and Sedat Han, "Using fuzzy risk assessment to rate cost overrun risk in international construction projects." *International Journal of Project Management*, Vol. 25 No. 5, July 2007, pp. 494-505.

20. Fa-Liang Gao, "A new way of predicting cement strength — Fuzzy logic". *Cement and Concrete Research*, Volume 27, Issue 6, June 1997, Pages 883–888.

21. Andrew Pollack, "Technology; Fuzzy Logic For Computers". *New York Times*, 11 October 1984; Andrew Pollack, "Fuzzy Computer Theory: How to Mimic the Mind?" *New York Times*, 2 April 1989.

22. The NAFIPS website URL is http://nafips.ece.ualberta.ca/

23. The IFSA URL is: http://isdlab.ie.ntnu.edu.tw/ntust/ifsa/

24. The EUSFLAT URL is: http://www.eusflat.org/.

25. Lotfi Zadeh, "Factual Information about the Impact of Fuzzy Logic". Berkeley Initiative in Soft Computing, at Electrical Engineering and Computer Sciences Department, University of Berkeley, California, circa 2014.

26. Andrei Popescu, "A general approach to fuzzy concepts". *Mathematical Logic Quarterly* Vol. 50, No. 3, 2005, pp. 265-280.

27. Radim Belohlavek and Vilem Vychodil, "What is a fuzzy concept lattice?" Department of Computer Science, Palacky University, Olomouc, 2005.

28. Daniel Kreiss, *Prototype Politics: Technology-Intensive Campaigning and the Data of Democracy*. Oxford University Press, 2016.

29. E.g. Mikael Collan,Mario Fedrizzi,Janusz Kacprzyk, *Fuzzy Technology: Present Applications and Future Challenges*. Heidelberg: Springer, 2016, p. 65f.; Daniel

J. Lewis and Trevor P. Martin, "Managing Vagueness with Fuzzy in Hierarchical Big Data". *Procedia Computer Science*, Volume 53, 2015, Pages 19-28.

30. Chris Preimesberger, "Big-Data Analytics Plays Big Role in 2016 Election Campaigns". *eWeek*, 24 September 2016. ;Gregory Thomas, "The Big Data Advantage in the Race for the White House." *Bemyapp Media*, 2 September 2016.; Alex Woodie, "Why Winning Politics Is Now Tied to Big Data Analytics". *Datanami.com*, 10 May 2016.; Lisa Ragusa, "And the Winner of the 2016 Election Is… Big Data". *Liaison*, 4 November 2016.; John Markman, "Big Data And The 2016 Election". *Forbes Magazine*, 8 August 2016.; Taylor Armerding, "Big Data and elections: The candidates know you – better than you know them." CSOonline.com, 17 July 2016.
31. Robert Epstein, "How Google Could Rig the 2016 Election". Politico.com, 19 August 2015 ; Marcel Rosenbach, "How Google and Facebook Can Reshape Elections", *Der Spiegel online* (English edition), 8 November 2016.
32. Preimesberger, ibid.
33. Kate Brannely, "Trump Campaign Pays Millions to Overseas Big Data Firm." NBC News, 4 November 2016.
34. Cambridge Analytica website.
35. Adam Tanner, "Nine Things You Don't Know About The Gathering Of Your Personal Data." *Forbes Magazine*, 4 November 2014.
36. Steve Lohr and Natasha Singernov, "How Data Failed Us in Calling an Election." *New York Times*, 10 November 2016
37. "How Trump won the presidency". Interview of Gerald F. Seib with Kellyanne Conway, *Wall Street Journal* (WSJ CEO Council full interview video), 14 November 2016. See also: Jonathan Vanian, "How Bad Polling Data Fooled Everyone Except Donald Trump". *Fortune*, 10 November 2016.
38. Ibid.
39. Ryan Lizza, "Kellyanne Conway's political machinations", *The New Yorker*, 17 October 2016.
40. Tableau.com, *Big data: the top 8 trends for 2016.*
41. Lotfi A. Zadeh, "Is there a need for fuzzy logic?", *Information Sciences*, No. 178, 2008, p. 2753.
42. Susan Haack, Philosophy of Logics. Cambridge University Press, 1978, p. 213.
43. Susan L. Epstein, "Memory and concepts in reactive learning". *Proceedings of the Canadian Workshop on Machine Learning 1992.*
44. Robert M. Wachter, "How Measurement Fails Doctors and Teachers". *New York Times*, 16 January 2016.
45. Matti Eklund, "Vagueness and Second-Level Indeterminacy", in: Richard Dietz & Sebastiano Moruzzi (eds.), *Cuts and clouds. Vagueness, Its Nature, and Its Logic.* Oxford University Press, 2009, p. 65.

46. Lotfi Zadeh, "What is fuzzy logic?". *IFSA Newsletter* (International Fuzzy Systems Association), Vol. 10, No. 1, March 2013, p. 5-6.

47. Alfred Korzybski, *Science and Sanity: An Introduction to Non-Aristotelian Systems and General Semantics* (5th ed.). Forest Hills, N.Y.: Institute of General Semantics, 1995. Gregory Bateson, *Steps to an Ecology of Mind: Collected Essays in Anthropology, Psychiatry, Evolution, and Epistemology*. Chicago: University Of Chicago Press, 1972.

48. Tanya Lewis, "What's the Universe Made Of? Math, Says Scientist." *Live Science*, 30 January 2014.

49. George Lakoff, "Hedges: A Study in Meaning Criteria and the Logic of Fuzzy Concepts." *Journal of Philosophical Logic*, Vol. 2, 1973, pp. 458-508.

50. Charles Ragin, *Redesigning Social Inquiry: Fuzzy Sets and Beyond*. University of Chicago Press, 2008. Shaomin Li, "Measuring the fuzziness of human thoughts: An application of fuzzy sets to sociological research". *The Journal of Mathematical Sociology*, Volume 14, Issue 1, 1989, pp. 67-84.

51. Ann Markusen, "Fuzzy Concepts, Scanty Evidence, Policy Distance: The Case for Rigour and Policy Relevance in Critical Regional Studies." In: *Regional Studies*, Volume 37, Issue 6-7, 2003, pp. 701-717.

52. Jörg Rössel and Randall Collins, "Conflict theory and interaction rituals. The microfoundations of conflict theory." In: Jonathan H. Turner (ed.), *Handbook of Sociological Theory*. New York: Springer, 2001, p. 527.

53. Loïc Wacquant, "The fuzzy logic of practical sense." in: Pierre Bourdieu and Loïc Wacquant, *An invitation to reflexive sociology*. London: Polity Press, 1992, chapter I section 4.

54. Ph. Manning "Fuzzy Description: Discovery and Invention in Sociology". In: *History of the Human Sciences*, Vol. 7, No. 1, 1994, pp. 117-23.

55. Philip Shenon, "Their prince is back: Cambodians are baffled." *New York Times*, 6 June 1993.

56. Betty Blair, "Interview with Lotfi Zadeh, Creator of Fuzzy Logic". *Azerbaijan International*, Winter 1994, pp. 46-47.

57. Johan van Benthem et al. (eds.), *The age of alternative logics. Assessing philosophy of logic and mathematics today*. Dordrecht: Springer, 2006, p. 203.

58. Masao Mukaidono, *Fuzzy logic for beginners*. Singapore: World Scientific Publishing, 2001.

59. Bertrand Russell, "Vagueness" (1923) in: *Bertrand Russell Papers*, Vol. 9, pp. 147–54. Nadine Faulkner, "Russell and vagueness." *Journal of Bertrand Russell Studies*, Summer 2003, pp. 43-63.

60. Patrick Hughes & George Brecht, *Vicious Circles and Infinity. An anthology of Paradoxes*. Penguin Books, 1978.

61. See further Radim Belohlavek & George J. Klir (eds.) *Concepts and Fuzzy Logic*. MIT Press, 2011. John R. Searle, "Minds, brains and programs". *The behavioral and brain sciences*, Vol. 3, No. 3, 1980, pp. 417-457.

62. Philip J. Kelman & Martha E. Arterberry, *The cradle of knowledge: development of perception in infancy*. Cambridge, Mass.: The MIT Press, 2000.

63. Bart Kosko, "Yes, Candidates, There Is a Fuzzy Math". *New York Times*, 7 November 2000.

64. David Henry, "Fuzzy Numbers', *Bloomberg Businessweek*, 3 October 2004.

65. Kazem Sadegh-Zadeh "The Fuzzy Revolution: Goodbye to the Aristotelian Weltanschauung". In: *Artificial Intelligence in Medicine*, 21, 2001, pp. 18-19.

66. Russel Gordon & David Bendien, "Standard classifications". *New Zealand Statistics Review*, September 1993, p. 20.

67. Ronald A. Havens (ed.), *The wisdom of Milton H. Erickson, Volume I: hypnosis and hypnotherapy*. New York: Irvington Publishers, 1992, p. 106. Joseph O'Connor & John Seymour (ed.), *Introducing neuro-linguistic programming*. London: Thorsons, 1995, p. 116f.

68. C.N. de Groot, "Sociology of religion looks at psychotherapy." *Recherches sociologiques* (Louvain-la-Neuve, Belgium), Vol. 29, No. 2, 1998, pp. 3-17 at p. 4.

69. Michael Hammond, Jane Howarth and Russell Keat, *Understanding Phenomenology*. Oxford: Blackwell, 1991.

70. Lotfi Zadeh, "What is fuzzy logic?". *IFSA Newsletter* (International Fuzzy Systems Association), Vol. 10, No. 1, March 2013. Takeshi Yamakawa "Stabilization of an Inverted Pendulum by a High-speed Fuzzy Logic Controller Hardware System". *Fuzzy Sets and Systems*, Vol.32, pp.161-180, 1989.

71. cf. Timothy Williamson, *Vagueness*. London: Routledge, 1996, p. 258.

CHAPTER 3

Fuzzy Set

Chapter Contents

- **Introduction**
- **Fuzzy Logic**
- **Fuzzy Interval**
- **Fuzzy Categories**
- **Entropy**
- **Extensions**
- **Fuzzy Set Operations**
- **Rough Fuzzy Hybridization**
- **Neuro-Fuzz**
- **Fuzzy Control System**
- **Fuzzy Finite Element**

Before illustrating the mechanisms which make fuzzy logic machines work, it is important to realize what fuzzy logic actually is. Fuzzy logic is a superset of conventional (Boolean) logic that has been extended to handle the concept of partial truth- truth values between "completely true" and "completely false". As its name suggests, it is the logic underlying modes of reasoning which are approximate rather than exact. The importance of fuzzy logic derives from the fact that most modes of human reasoning and especially common sense reasoning are approximate in nature.

The essential characteristics of fuzzy logic as founded by Zader Lotfi are as follows.

- In fuzzy logic, exact reasoning is viewed as a limiting case of approximate reasoning.
- In fuzzy logic everything is a matter of degree.
- Any logical system can be fuzzified

- In fuzzy logic, knowledge is interpreted as a collection of elastic or, equivalently , fuzzy constraint on a collection of variables
- Inference is viewed as a process of propagation of elastic constraints.

The third statement hence, define Boolean logic as a subset of Fuzzy logic.

INTRODUCTION

In mathematics, **fuzzy sets** are sets whose elements have degrees of membership. Fuzzy sets were introduced by Lotfi A. Zadeh and Dieter Klaua in 1965 as an extension of the classical notion of set. At the same time, Salii (1965) defined a more general kind of structure called an L-relation, which he studied in an abstract algebraic context. Fuzzy relations, which are used now in different areas, such as linguistics (De Cock, Bodenhofer & Kerre 2000) decision-making (Kuzmin 1982) and clustering (Bezdek 1978), are special cases of L-relations when L is the unit interval [0, 1]. In classical set theory, the membership of elements in a set is assessed in binary terms according to a bivalent condition — an element either belongs or does not belong to the set. By contrast, fuzzy set theory permits the gradual assessment of the membership of elements in a set; this is described with the aid of a membership function valued in the real unit interval [0, 1]. Fuzzy sets generalize classical sets, since the indicator functions of classical sets are special cases of the membership functions of fuzzy sets, if the latter only take values 0 or 1. In fuzzy set theory, classical bivalent sets are usually called *crisp* sets. The fuzzy set theory can be used in a wide range of domains in which information is incomplete or imprecise, such as bioinformatics.

A fuzzy set is a pair (U, m) where U is a set and $m: U \to [0, 1]$ a membership function.

For each $x \in U$, the value $m(x)$ is called the **grade** of membership of x in (U, m). For a finite set $U = \{x_1, \ldots, x_n\}$, the fuzzy set (U, m) is often denoted by $\{m(x_1)/x_1, \ldots, m(x_n)/x_n\}$.

Let $x \in U$. Then x is called **not included** in the fuzzy set (U, m) if $m(x) = 0$, x is called **fully included** if $m(x) = 1$, and x is called a **fuzzy member** if $0 < m(x) < 1$. The set $\{x \in U \mid m(x) > 0\}$ is called the **support** of (U, m) and the set $\{x \in U \mid m(x) = 1\}$ is called its **kernel** or **core**. The function m is called the **membership function** of the fuzzy set (U, m).

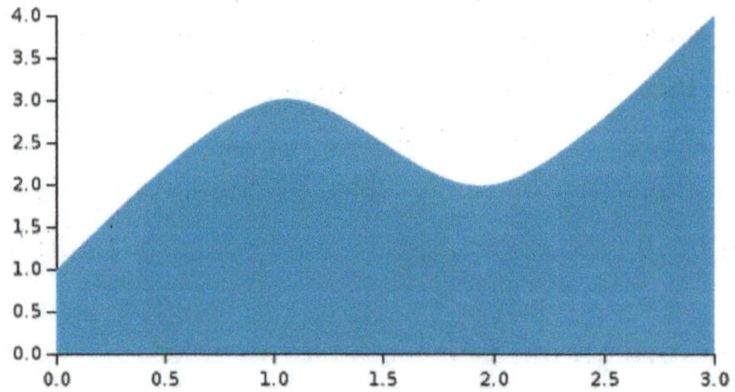

Sometimes, more general variants of the notion of fuzzy set are used, with membership functions taking values in a (fixed or variable) algebra or structure L of a given kind; usually it is required that L be at least a poset or lattice. These are usually called ***L*-fuzzy sets**, to distinguish them from those valued over the unit interval. The usual membership functions with values in [0, 1] are then called [0, 1]-valued membership functions. These kinds of generalizations were first considered in 1967 by Joseph Goguen, who was a student of Zadeh.

FUZZY LOGIC

As an extension of the case of multi-valued logic, valuations $(\mu : V_o \to W)$ of propositional variables (V_o) into a set of membership degrees (W) can be thought of as membership functions mapping predicates into fuzzy sets (or more formally, into an ordered set of fuzzy pairs, called a fuzzy relation). With these valuations, many-valued logic can be extended to allow for fuzzy premises from which graded conclusions may be drawn.

This extension is sometimes called "fuzzy logic in the narrow sense" as opposed to "fuzzy logic in the wider sense," which originated in the engineering fields of automated control and knowledge engineering, and which encompasses many topics involving fuzzy sets and "approximated reasoning."

Industrial applications of fuzzy sets in the context of "fuzzy logic in the wider sense" can be found at fuzzy logic.

Fuzzy Number

A **fuzzy number** is a convex, normalized fuzzy set $\tilde{A} \subseteq \mathbb{R}$ whose membership function is at least segmentally continuous and has the functional value $\mu_A(x) = 1$ at at least one element.

This can be likened to the funfair game "guess your weight," where someone guesses the contestant's weight, with closer guesses being more correct, and where the guesser "wins" if he or she guesses near enough to the contestant's weight, with the actual weight being completely correct (mapping to 1 by the membership function).

FUZZY INTERVAL

A **fuzzy interval** is an uncertain set $\tilde{A} \subseteq \mathbb{R}$ with a mean interval whose elements possess the membership function value $\mu_A(x) = 1$. As in fuzzy numbers, the membership function must be convex, normalized, at least segmentally continuous.

FUZZY CATEGORIES

The use of set membership as a key components of category theory can be generalized to fuzzy sets. This approach which initiated in 1968 shortly after the introduction of fuzzy set theory led to the development of "Goguen categories" in the 21st century. In these categories, rather than using two valued set membership, more general intervals are used, and may be lattices as in L-fuzzy sets.

FUZZY RELATION EQUATION

The fuzzy relation equation is an equation of the form A · R = B, where A and B are fuzzy sets, R is a fuzzy relation, and A · R stands for the composition of A with R.

ENTROPY

Let A be a fuzzy variable with a continuous membership function. Then its entropy is

$$H[A] = \int_{-\infty}^{\infty} S(\mathrm{Cr}\{A \geq t\})\, dt$$

where

$$S(y) = -y \ln y - (1-y)\ln(1-y).$$

EXTENSIONS

There are many mathematical constructions similar to or more general than fuzzy sets. Since fuzzy sets were introduced in 1965, a lot of new mathematical constructions and theories treating imprecision, inexactness, ambiguity, and uncertainty have been developed. Some of these constructions and theories are extensions of fuzzy set theory, while others try to mathematically model imprecision and uncertainty in a different way (Burgin & Chunihin 1997; Kerre 2001; Deschrijver and Kerre, 2003).

The diversity of such constructions and corresponding theories includes:

- interval sets (Moore, 1966),
- *L*-fuzzy sets (Goguen, 1967),
- flou sets (Gentilhomme, 1968),
- Boolean-valued fuzzy sets (Brown, 1971),
- type-2 fuzzy sets and type-n fuzzy sets (Zadeh, 1975),
- set-valued sets (Chapin, 1974; 1975),
- interval-valued fuzzy sets (Grattan-Guinness, 1975; Jahn, 1975; Sambuc, 1975; Zadeh, 1975),
- functions as generalizations of fuzzy sets and multisets (Lake, 1976),
- level fuzzy sets (Radecki, 1977)
- underdetermined sets (Narinyani, 1980),
- rough sets (Pawlak, 1982),
- intuitionistic fuzzy sets (Atanassov, 1983),
- fuzzy multisets (Yager, 1986),

- intuitionistic L-fuzzy sets (Atanassov, 1986),
- rough multisets (Grzymala-Busse, 1987),
- fuzzy rough sets (Nakamura, 1988),
- real-valued fuzzy sets (Blizard, 1989),
- named sets (Burgin, 1990),
- vague sets (Wen-Lung Gau and Buehrer, 1993),
- Q-sets (Gylys, 1994)
- shadowed sets (Pedrycz, 1998),
- α-level sets (Yao, 1997),
- genuine sets (Demirci, 1999),
- soft sets (Molodtsov, 1999),
- intuitionistic fuzzy rough sets (Cornelis, De Cock and Kerre, 2003)
- blurry sets (Smith, 2004)
- L-fuzzy rough sets (Radzikowska and Kerre, 2004),
- generalized rough fuzzy sets (Feng, 2010)
- rough intuitionistic fuzzy sets (Thomas and Nair, 2011),
- soft rough fuzzy sets (Meng, Zhang and Qin, 2011)
- soft fuzzy rough sets (Meng, Zhang and Qin, 2011)
- soft multisets (Alkhazaleh, Salleh and Hassan, 2011)
- fuzzy soft multisets (Alkhazaleh and Salleh, 2012)
- bipolar fuzzy sets (Wen-Ran Zhang, 1998)

While most of the above can be generally categorized as truth-based extensions to fuzzy sets, bipolar fuzzy set theory presents a philosophically and logically different, equilibrium-based generalization of fuzzy sets.

FUZZY SET OPERATIONS

A **fuzzy set operation** is an operation on fuzzy sets. These operations are generalization of crisp set operations. There is more than one possible generalization. The most widely used operations are called *standard fuzzy set operations*. There are three operations: fuzzy complements, fuzzy intersections, and fuzzy unions.

Standard Fuzzy Set Operations

Let A and B be fuzzy sets that A,B ∈ U, u is an element in the U universe (e.g. value)

Standard complement

$\neg\mu_A(u) = 1 - \mu_A(u)$

Standard Intersection

$$\mu_{A\cap B}(u) = \min\{\mu_A(u), \mu_B(u)\}$$

Standard union

$$\mu_{A\cup B}(u) = max\{\mu_A(u), \mu_B(u)\}$$

Fuzzy Complements

$A(x)$ is defined as the degree to which x belongs to A. Let cA denote a fuzzy complement of A of type c. Then $cA(x)$ is the degree to which x belongs to cA, and the degree to which x does not belong to A. ($A(x)$ is therefore the degree to which x does not belong to cA.) Let a complement cA be defined by a function

$c : [0,1] \rightarrow [0,1]$

$c(A(x)) = cA(x)$

Axioms for fuzzy complements

Axiom c1. ***Boundary condition***

$c(0) = 1$ and $c(1) = 0$

Axiom c2. ***Monotonicity***

For all a, $b \in [0, 1]$, if $a < b$, then $c(a) > c(b)$

Axiom c3. ***Continuity***

c is continuous function.

Axiom c4. ***Involutions***

c is an involution, which means that $c(c(a)) = a$ for each $a \in [0,1]$

Fuzzy Intersections

The intersection of two fuzzy sets A and B is specified in general by a binary operation on the unit interval, a function of the form

i:$[0,1]\times[0,1] \rightarrow [0,1]$.

$(A \cap B)(x) = i[A(x), B(x)]$ for all x.

Axioms for Fuzzy Intersection

Axiom i1. ***Boundary condition***

$i(a, 1) = a$

Axiom i2. ***Monotonicity***

$b \leq d$ implies $i(a,b) \leq i(a,d)$

Axiom i3. ***Commutativity***

$i(a, b) = i(b, a)$

Axiom i4. ***Associativity***

$i(a, i(b, d)) = i(i(a, b), d)$

Axiom i5. ***Continuity***

i is a continuous function

Axiom i6. ***Subidempotency***

$i(a, a) \leq a$

Fuzzy Unions

The union of two fuzzy sets A and B is specified in general by a binary operation on the unit interval function of the form

u:[0,1]×[0,1] → [0,1].

$(A \cup B)(x) = u[A(x), B(x)]$ for all x

Axioms for Fuzzy Union

Axiom u1. ***Boundary condition***

$u(a, 0) = u(0, a) = a$

Axiom u2. ***Monotonicity***

$b \leq d$ implies $u(a, b) \leq u(a, d)$

Axiom u3. ***Commutativity***

$u(a, b) = u(b, a)$

Axiom u4. ***Associativity***

$u(a, u(b, d)) = u(u(a, b), d)$

Axiom u5. ***Continuity***

u is a continuous function

Axiom u6. ***Superidempotency***

$u(a, a) \geq a$

Axiom u7. ***Strict monotonicity***

$a_1 < a_2$ and $b_1 < b_2$ implies $u(a_1, b_1) < u(a_2, b_2)$

Aggregation operations

Aggregation operations on fuzzy sets are operations by which several fuzzy sets are combined in a desirable way to produce a single fuzzy set.

Aggregation operation on n fuzzy set ($2 \leq n$) is defined by a function

$h:[0,1]^n \rightarrow [0,1]$

Axioms for aggregation operations fuzzy sets

Axiom h1. ***Boundary condition***

$h(0, 0, ..., 0) = 0$ and $h(1, 1, ..., 1) = 1$

Axiom h2. ***Monotonicity***

For any pair $<a_1, a_2, ..., a_n>$ and $<b_1, b_2, ..., b_n>$ of n-tuples such that $a_i, b_i \in [0,1]$ for all $i \in N_n$, if $a_i \leq b_i$ for all $i \in N_n$, then $h(a_1, a_2, ..., a_n) \leq h(b_1, b_2, ..., b_n)$; that is, h is monotonic increasing in all its arguments.

Axiom h3. ***Continuity***

h is a continuous function.

ROUGH FUZZY HYBRIDIZATION

Rough fuzzy hybridization is a method of hybrid intelligent system or soft computing, where Fuzzy set theory is used for linguistic representation of patterns, leading to a *fuzzy granulation* of the feature space. Rough set theory is used to obtain dependency rules which model informative regions in the granulated feature space.

Type-2 Fuzzy Sets and Systems

Type-2 fuzzy sets and systems generalize Type-1 fuzzy sets and systems so that more uncertainty can be handled. From the very beginning of fuzzy sets, criticism was made about the fact that the membership function of a type-1 fuzzy set has no uncertainty associated with it, something that seems to contradict the word *fuzzy*, since that word has the connotation of lots of uncertainty. So, what does one do when there is uncertainty about the value of the membership function? The answer to this question was provided in 1975 by the inventor of fuzzy sets, Prof. Lotfi A. Zadeh , when he proposed more sophisticated kinds of fuzzy sets, the first of which he called a ***type-2 fuzzy set***. A type-2 fuzzy set lets us incorporate uncertainty about the membership function into fuzzy set theory, and is a way to address the above criticism of type-1 fuzzy sets head-on. And, if there is no uncertainty, then a type-2 fuzzy set reduces to a type-1 fuzzy set, which is analogous to probability reducing to determinism when unpredictability vanishes,.

In order to symbolically distinguish between a type-1 fuzzy set and a type-2 fuzzy set, a tilde symbol is put over the symbol for the fuzzy set; so, A denotes a type-1 fuzzy set, whereas Ã denotes the comparable type-2 fuzzy set. When the latter is done, the resulting type-2 fuzzy set is called a ***general type-2 fuzzy set*** (to distinguish it from the special interval type-2 fuzzy set).

Prof. Zadeh didn"t stop with type-2 fuzzy sets, because in that 1976 paper he also generalized all of this to type-*n* fuzzy sets. The present article focuses only on type-2 fuzzy sets because they are the *next step* in the logical progression from type-1 to type-*n* fuzzy sets, where $n = 1, 2, \ldots$. Although some researchers are beginning to explore higher than type-2 fuzzy sets, as of early 2009, this work is in its infancy.

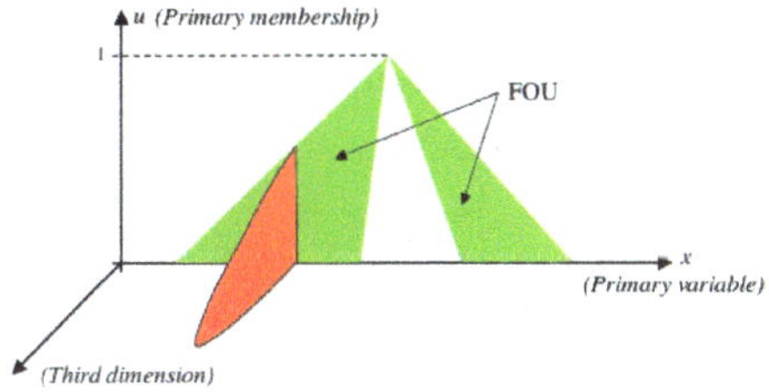

Figure 1: The membership function of a general type-2 fuzzy set is three-dimensional. A cross-section of one slice of the third dimension is shown. This cross-section, as well as all others, sits on the FOU. Only the boundary of the cross-section is used to describe the membership function of a general type-2 fuzzy set. It is shown filled-in for artistic purposes.

The membership function of a general type-2 fuzzy set, Ã, is three-dimensional (Fig. 1), where the third dimension is the value of the membership function at each point on its two-dimensional domain that is called its *footprint of uncertainty* (FOU).

For an interval type-2 fuzzy set that third-dimension value is the same (e.g., 1) everywhere, which means that no new information is contained in the third dimension of an interval type-2 fuzzy set. So, for such a set, the third dimension is ignored, and only the FOU is used to describe it. It is for this reason that an interval type-2 fuzzy set is sometimes called a *first-order uncertainty* fuzzy set model, whereas a general type-2 fuzzy set (with its useful third-dimension) is sometimes referred to as a *second-order uncertainty* fuzzy set model.

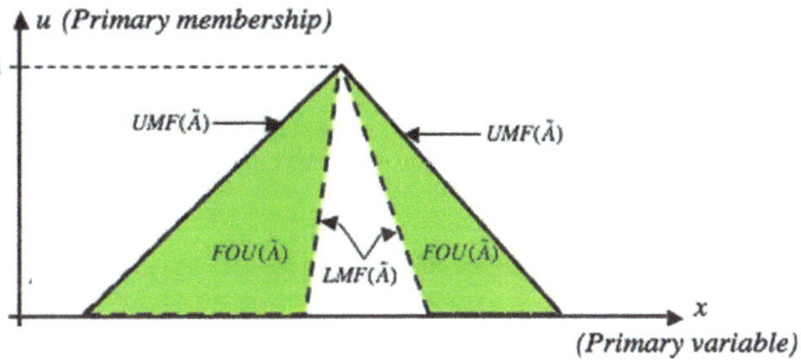

Figure 2: FOU for an interval type-2 fuzzy set. Many other shapes are possible for the FOU.

The FOU represents the blurring of a type-1 membership function, and is completely described by its two bounding functions (Fig. 2), a lower membership function (LMF) and an upper membership function (UMF), both of which are type-1 fuzzy sets! Consequently, it is possible to use type-1 fuzzy set mathematics to characterize and work with interval type-2 fuzzy sets. This means that engineers and scientists who already know type-1 fuzzy sets will not have to invest a lot of time learning about general type-2 fuzzy set mathematics in order to understand and use interval type-2 fuzzy sets.

Work on type-2 fuzzy sets languished during the 1980s and early-to-mid 1990's, although a small number of articles were published about them. People were still trying to figure out what to do with type-1 fuzzy sets, so even though Zadeh proposed type-2 fuzzy sets in 1976, the time was not right for researchers to drop what they were doing with type-1 fuzzy sets to focus on type-2 fuzzy sets. This changed in the latter part of the 1990s as a result of Prof. Jerry Mendel and his student's works on type-2 fuzzy sets and systems (e.g.,). Since then, more and more researchers around the world are writing articles about type-2 fuzzy sets and systems.

Interval Type-2 Fuzzy Sets

Interval type-2 fuzzy sets have received the most attention because the mathematics that is needed for such sets—primarily Interval arithmetic—is much simpler than the mathematics that is needed for general type-2 fuzzy sets. So, the literature about interval type-2 fuzzy sets is large, whereas the literature about general type-2 fuzzy sets is much smaller. Both kinds of fuzzy sets are being actively researched by an ever-growing number of researchers around the world.

Formilleri for the following have already been worked out for interval type-2 fuzzy sets:

- Fuzzy set operations: union, intersection and complement (,)
- Centroid (a very widely used operation by practitioners of such sets, and also an important uncertainty measure for them) (, ,)
- Other uncertainty measures [fuzziness, cardinality, variance and skewness and uncertainty bounds
- Similarity (, ,)
- Subsethood
- Embedded fuzzy sets (, ,)
- Fuzzy set ranking
- Fuzzy rule ranking and selection
- Type-reduction methods (,)
- Firing intervals for an interval type-2 fuzzy logic system (, ,)
- Fuzzy weighted average
- Linguistic weighted average
- Synthesizing an FOU from data that are collected from a group of subject

Interval Type-2 Fuzzy Logic Systems

Type-2 fuzzy sets are finding very wide applicability in ***rule-based fuzzy logic systems*** (FLSs) because they let uncertainties be modeled by them whereas such uncertainties cannot be modeled by type-1 fuzzy sets. A block diagram of a type-2 FLS is depicted in Fig. 3. This kind of FLS is used in fuzzy logic control, fuzzy logic signal processing, rule-based classification, etc., and is sometimes referred to as a *function approximation* application of fuzzy sets, because the FLS is designed to minimize an error function.

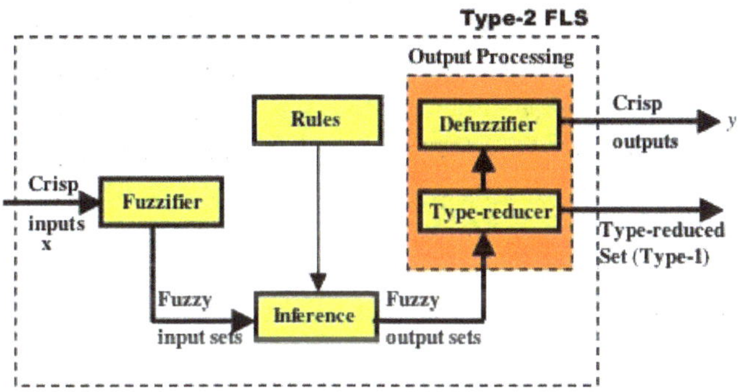

Figure 3: Type-2 FLS

The following discussions, about the four components in the Fig. 3 rule-based FLS, are given for an interval type-2 FLS, because to-date they are the most popular kind of type-2 FLS; however, most of the discussions are also applicable for a general type-2 FLS.

Rules, that are either provided by subject experts or are extracted from numerical data, are expressed as a collection of IF-THEN statements, e.g.,

IF temperature is *moderate* and pressure is *high*, then rotate the valve *a bit to the right*.

Fuzzy sets are associated with the terms that appear in the antecedents (IF-part) or consequents (THEN-part) of rules, and with the inputs to and the outputs of the FLS. Membership functions are used to describe these fuzzy sets, and in a type-1 FLS they are all type-1 fuzzy sets, whereas in an interval type-2 FLS at least one membership function is an interval type-2 fuzzy set.

An interval type-2 FLS lets any one or all of the following kinds of uncertainties be quantified:

- Words that are used in antecedents and consequents of rules—because words can mean different things to different people.
- Uncertain consequents—because when rules are obtained from a group of experts, consequents will often be different for the same rule, i.e. the experts will not necessarily be in agreement.
- Membership function parameters—because when those parameters are optimized using uncertain (noisy) training data, the parameters become uncertain.
- Noisy measurements—because very often it is such measurements that activate the FLS.

In Fig. 3, measured (crisp) inputs are first transformed into fuzzy sets in the ***Fuzzifier*** block because it is fuzzy sets and not numbers that activate the rules which are described in terms of fuzzy sets and not numbers. Three kinds of fuzzifiers are possible in an interval type-2 FLS. When measurements are:

- Perfect, they are modeled as a crisp set;
- Noisy, but the noise is stationary, they are modeled as a type-1 fuzzy set; and,

- Noisy, but the noise is non-stationary, they are modeled as an interval type-2 fuzzy set (this latter kind of fuzzification cannot be done in a type-1 FLS).

In Fig. 3, after measurements are fuzzified, the resulting input fuzzy sets are mapped into fuzzy output sets by the ***Inference*** block. This is accomplished by first quantifying each rule using fuzzy set theory, and by then using the mathematics of fuzzy sets to establish the output of each rule, with the help of an inference mechanism. If there are M rules then the fuzzy input sets to the Inference block will activate only a subset of those rules, where the subset contains at least one rule and usually way fewer than M rules. Inference is done one rule at a time. So, at the output of the Inference block, there will be one or more *fired-rule fuzzy output sets*.

In most engineering applications of a FLS, a number (and not a fuzzy set) is needed as its final output, e.g., the consequent of the rule given above is "Rotate the valve a bit to the right." No automatic valve will know what this means because "a bit to the right" is a linguistic expression, and a valve must be turned by numerical values, i.e. by a certain number of degrees. Consequently, the fired-rule output fuzzy sets have to be converted into a number, and this is done in the Fig. 3 ***Output Processing*** block.

In a type-1 FLS, output processing, called ***Defuzzification***, maps a type-1 fuzzy set into a number. There are many ways for doing this, e.g., compute the union of the fired-rule output fuzzy sets (the result is another type-1 fuzzy set) and then compute the center of gravity of the membership function for that set; compute a weighted average of the center of gravities of each of the fired rule consequent membership functions; etc.

Things are somewhat more complicated for an interval type-2 FLS, because to go from an interval type-2 fuzzy set to a number (usually) requires two steps (Fig. 3). The first step, called ***type-reduction***, is where an interval type-2 fuzzy set is reduced to an interval-valued type-1 fuzzy set. There are as many type-reduction methods as there are type-1 defuzzification methods. An algorithm developed by Karnik and Mendel (,) now known as the ***KM Algorithm*** is used for type-reduction. Although this algorithm is iterative, it is very fast.

The second step of Output Processing, which occurs after type-reduction, is still called ***defuzzification***. Because a type-reduced set of an interval type-2 fuzzy set is always a finite interval of numbers, the defuzzified value is just the average of the two end-points of this interval.

It is clear from Fig. 3 that there can be two outputs to an interval type-2 FLS—crisp numerical values and the type-reduced set. The latter provides a measure of the uncertainties that have flowed through the interval type-2 FLS, due to the (possibly) uncertain input measurements that have activated rules whose antecedents or consequents or both are uncertain. Just as standard deviation is widely used in probability and statistics to provide a measure of unpredictable uncertainty about a mean value, the type-reduced set can provided a measure of uncertainty about the crisp output of an interval type-2 FLS.

Computing with Words

Another application for fuzzy sets has also been inspired by Prof. Zadeh (–)—***Computing With Words***. Different acronyms have been used for "computing with words," e.g., CW and CWW. According to Zadeh:

CWW is a methodology in which the objects of computation are words and propositions drawn from a natural language. [It is] inspired by the remarkable human capability to perform a wide variety of physical and mental tasks without any measurements and any computations.

Of course, he did not mean that computers would actually compute using words—single words or phrases—rather than numbers. He meant that computers would be activated by words, which would be converted into a mathematical representation using fuzzy sets and that these fuzzy sets would be mapped by a CWW engine into some other fuzzy set after which the latter would be converted back into a word. A natural question to ask is: Which kind of fuzzy set—type-1 or type-2—should be used as a model for a word? Mendel (,) has argued, on the basis of Karl Popper's concept of ***Falsificationism*** (,), that using a type-1 fuzzy set as a model for a word is scientifically incorrect. An interval type-2 fuzzy set should be used as a (first-order uncertainty) model for a word. Much research is under way about CWW.

Applications

Type-2 fuzzy sets were applied in image processing, video processing and computer vision.

NEURO-FUZZY

In the field of artificial intelligence, **neuro-fuzzy** refers to combinations of artificial neural networks and fuzzy logic. Neuro-fuzzy was proposed by J. S. R. Jang. Neuro-fuzzy hybridization results in a hybrid intelligent system that synergizes these two techniques by combining the human-like reasoning style of fuzzy systems with the learning and connectionist structure of neural networks. Neuro-fuzzy hybridization is widely termed as Fuzzy Neural Network (FNN) or Neuro-Fuzzy System (NFS) in the literature. Neuro-fuzzy system (the more popular term is used henceforth) incorporates the human-like reasoning style of fuzzy systems through the use of fuzzy sets and a linguistic model consisting of a set of IF-THEN fuzzy rules. The main strength of neuro-fuzzy systems is that they are universal approximators with the ability to solicit interpretable IF-THEN rules.

The strength of neuro-fuzzy systems involves two contradictory requirements in fuzzy modeling: interpretability versus accuracy. In practice, one of the two properties prevails. The neuro-fuzzy in fuzzy modeling research field is divided into two areas: linguistic fuzzy modeling that is focused on interpretability, mainly the Mamdani model; and precise fuzzy modeling that is focused on accuracy, mainly the Takagi-Sugeno-Kang (TSK) model.

Although generally assumed to be the realization of a fuzzy system through connectionist networks, this term is also used to describe some other configurations including:

- Deriving fuzzy rules from trained RBF networks.
- Fuzzy logic based tuning of neural network training parameters.
- Fuzzy logic criteria for increasing a network size.
- Realising fuzzy membership function through clustering algorithms in unsupervised learning in SOMs and neural networks.

- Representing fuzzification, fuzzy inference and defuzzification through multi-layers feed-forward connectionist networks.

It must be pointed out that interpretability of the Mamdani-type neuro-fuzzy systems can be lost. To improve the interpretability of neuro-fuzzy systems, certain measures must be taken, wherein important aspects of interpretability of neuro-fuzzy systems are also discussed.

A recent research line addresses the data stream mining case, where neuro-fuzzy systems are sequentially updated with new incoming samples on demand and on-the-fly. Thereby, system updates do not only include a recursive adaptation of model parameters, but also a dynamic evolution and pruning of model components (neurons, rules), in order to handle concept drift and dynamically changing system behavior adequately and to keep the systems/models "up-to-date" anytime. Comprehensive surveys of various evolving neuro-fuzzy systems approaches can be found in and.

Pseudo outer-product-based Fuzzy Neural Networks

Pseudo outer-product-based fuzzy neural networks ("POPFNN") are a family of neuro-fuzzy systems that are based on the linguistic fuzzy model.

Three members of POPFNN exist in the literature:

- **POPFNN-AARS(S)**, which is based on the Approximate Analogical Reasoning Scheme
- **POPFNN-CRI(S)**, which is based on commonly accepted fuzzy Compositional Rule of Inference
- **POPFNN-TVR**, which is based on Truth Value Restriction

The "POPFNN" architecture is a five-layer neural network where the layers from 1 to 5 are called: input linguistic layer, condition layer, rule layer, consequent layer, output linguistic layer. The fuzzification of the inputs and the defuzzification of the outputs are respectively performed by the input linguistic and output linguistic layers while the fuzzy inference is collectively performed by the rule, condition and consequence layers.

The learning process of POPFNN consists of three phases:

- Fuzzy membership generation
- Fuzzy rule identification
- Supervised fine-tuning

Various fuzzy membership generation algorithms can be used: Learning Vector Quantization (LVQ), Fuzzy Kohonen Partitioning (FKP) or Discrete Incremental Clustering (DIC). Generally, the POP algorithm and its variant LazyPOP are used to identify the fuzzy rules.

FUZZY CONTROL SYSTEM

A **fuzzy control system** is a control system based on fuzzy logic—a mathematical system that analyzes analog input values in terms of logical variables that take on continuous values between 0 and 1, in contrast to classical or digital logic, which operates on discrete values of either 1 or 0 (true or false, respectively).

Fuzzy logic is widely used in machine control. The term "fuzzy" refers to the fact that the logic involved can deal with concepts that cannot be expressed as the "true" or "false" but rather as "partially true". Although alternative approaches such as genetic algorithms and neural networks can perform just as well as fuzzy logic in many cases, fuzzy logic has the advantage that the solution to the problem can be cast in terms that human operators can understand, so that their experience can be used in the design of the controller. This makes it easier to mechanize tasks that are already successfully performed by humans.

History and Applications

Fuzzy logic was first proposed by Lotfi A. Zadeh of the University of California at Berkeley in a 1965 paper. He elaborated on his ideas in a 1973 paper that introduced the concept of "linguistic variables", which in this article equates to a variable defined as a fuzzy set. Other research followed, with the first industrial application, a cement kiln built in Denmark, coming on line in 1975.

Fuzzy systems were initially implemented in Japan.

- Interest in fuzzy systems was sparked by Seiji Yasunobu and Soji Miyamoto of Hitachi, who in 1985 provided simulations that demonstrated the feasibility of fuzzy control systems for the Sendai railway. Their ideas were adopted, and fuzzy systems were used to control accelerating, braking, and stopping when the line opened in 1987.
- In 1987, Takeshi Yamakawa demonstrated the use of fuzzy control, through a set of simple dedicated fuzzy logic chips, in an "inverted pendulum" experiment. This is a classic control problem, in which a vehicle tries to keep a pole mounted on its top by a hinge upright by moving back and forth. Yamakawa subsequently made the demonstration more sophisticated by mounting a wine glass containing water and even a live mouse to the top of the pendulum: the system maintained stability in both cases. Yamakawa eventually went on to organize his own fuzzy-systems research lab to help exploit his patents in the field.
- Japanese engineers subsequently developed a wide range of fuzzy systems for both industrial and consumer applications. In 1988 Japan established the Laboratory for International Fuzzy Engineering (LIFE), a cooperative arrangement between 48 companies to pursue fuzzy research. The automotive company Volkswagen was the only foreign corporate member of LIFE, dispatching a researcher for a duration of three years.
- Japanese consumer goods often incorporate fuzzy systems. Matsushita vacuum cleaners use microcontrollers running fuzzy algorithms to interrogate dust sensors and adjust suction power accordingly. Hitachi washing machines use fuzzy controllers to load-weight, fabric-mix, and dirt sensors and automatically set the wash cycle for the best use of power, water, and detergent.
- Canon developed an autofocusing camera that uses a charge-coupled device (CCD) to measure the clarity of the image in six regions of its field of view and use the information provided to determine if the image is in focus. It also tracks the rate of change of lens movement during focusing, and controls its speed to prevent overshoot. The camera's fuzzy control system uses 12 inputs: 6 to

obtain the current clarity data provided by the CCD and 6 to measure the rate of change of lens movement. The output is the position of the lens. The fuzzy control system uses 13 rules and requires 1.1 kilobytes of memory.

- An industrial air conditioner designed by Mitsubishi uses 25 heating rules and 25 cooling rules. A temperature sensor provides input, with control outputs fed to an inverter, a compressor valve, and a fan motor. Compared to the previous design, the fuzzy controller heats and cools five times faster, reduces power consumption by 24%, increases temperature stability by a factor of two, and uses fewer sensors.
- Other applications investigated or implemented include: character and handwriting recognition; optical fuzzy systems; robots, including one for making Japanese flower arrangements; voice-controlled robot helicopters (hovering is a "balancing act" rather similar to the inverted pendulum problem); rehabilitation robotics to provide patient-specific solutions (e.g. to control heart rate and blood pressure); control of flow of powders in film manufacture; elevator systems; and so on.

Work on fuzzy systems is also proceeding in the United State and Europe, although on a less extensive scale than in Japan.

- The US Environmental Protection Agency has investigated fuzzy control for energy-efficient motors, and NASA has studied fuzzy control for automated space docking: simulations show that a fuzzy control system can greatly reduce fuel consumption.
- Firms such as Boeing, General Motors, Allen-Bradley, Chrysler, Eaton, and Whirlpool have worked on fuzzy logic for use in low-power refrigerators, improved automotive transmissions, and energy-efficient electric motors.
- In 1995 Maytag introduced an "intelligent" dishwasher based on a fuzzy controller and a "one-stop sensing module" that combines a thermistor, for temperature measurement; a conductivity sensor, to measure detergent level from the ions present in the wash; a turbidity sensor that measures scattered and transmitted light to measure the soiling of the wash; and a magnetostrictive sensor to read spin rate. The system determines the optimum wash cycle for any load to obtain the best results with the least amount of energy, detergent, and water. It even adjusts for dried-on foods by tracking the last time the door was opened, and estimates the number of dishes by the number of times the door was opened.

Research and development is also continuing on fuzzy applications in software, as opposed to firmware, design, including fuzzy expert systems and integration of fuzzy logic with neural-network and so-called adaptive "genetic" software systems, with the ultimate goal of building "self-learning" fuzzy-control systems. These systems can be employed to control complex, nonlinear dynamic plants, for example, human body.

Fuzzy Sets

The input variables in a fuzzy control system are in general mapped by sets of membership functions similar to this, known as "fuzzy sets". The process of converting a crisp input value to a fuzzy value is called "fuzzification".

A control system may also have various types of switch, or "ON-OFF", inputs along with its analog inputs, and such switch inputs of course will always have a truth value equal to either 1 or 0, but the scheme can deal with them as simplified fuzzy functions that happen to be either one value or another.

Given "mappings" of input variables into membership functions and truth values, the microcontroller then makes decisions for what action to take, based on a set of "rules", each of the form:

IF brake temperature IS warm AND speed IS not very fast

THEN brake pressure IS slightly decreased.

In this example, the two input variables are "brake temperature" and "speed" that have values defined as fuzzy sets. The output variable, "brake pressure" is also defined by a fuzzy set that can have values like "static" or "slightly increased" or "slightly decreased" etc.

This rule by itself is very puzzling since it looks like it could be used without bothering with fuzzy logic, but remember that the decision is based on a set of rules:

- All the rules that apply are invoked, using the membership functions and truth values obtained from the inputs, to determine the result of the rule.
- This result in turn will be mapped into a membership function and truth value controlling the output variable.
- These results are combined to give a specific ("crisp") answer, the actual brake pressure, a procedure known as "defuzzification".

This combination of fuzzy operations and rule-based "inference" describes a "fuzzy expert system".

Traditional control systems are based on mathematical models in which the control system is described using one or more differential equations that define the system response to its inputs. Such systems are often implemented as "PID controllers" (proportional-integral-derivative controllers). They are the products of decades of development and theoretical analysis, and are highly effective.

If PID and other traditional control systems are so well-developed, why bother with fuzzy control? It has some advantages. In many cases, the mathematical model of the control process may not exist, or may be too "expensive" in terms of computer processing power and memory, and a system based on empirical rules may be more effective.

Furthermore, fuzzy logic is well suited to low-cost implementations based on cheap sensors, low-resolution analog-to-digital converters, and 4-bit or 8-bit one-chip microcontroller chips. Such systems can be easily upgraded by adding new rules to improve performance or add new features. In many cases, fuzzy control can be used to improve existing traditional controller systems by adding an extra layer of intelligence to the current control method.

Fuzzy Control in Detail

Fuzzy controllers are very simple conceptually. They consist of an input stage, a processing stage, and an output stage. The input stage maps sensor or other inputs, such as switches, thumbwheels, and so on, to the appropriate membership functions and truth values. The processing stage invokes each appropriate rule and generates a result for each, then combines the results of the rules. Finally, the output stage converts the combined result back into a specific control output value.

The most common shape of membership functions is triangular, although trapezoidal and bell curves are also used, but the shape is generally less important than the number of curves and their placement. From three to seven curves are generally appropriate to cover the required range of an input value, or the "universe of discourse" in fuzzy jargon.

As discussed earlier, the processing stage is based on a collection of logic rules in the form of IF-THEN statements, where the IF part is called the "antecedent" and the THEN part is called the "consequent". Typical fuzzy control systems have dozens of rules.

Consider a rule for a thermostat:

IF (temperature is "cold") THEN (heater is "high")

This rule uses the truth value of the "temperature" input, which is some truth value of "cold", to generate a result in the fuzzy set for the "heater" output, which is some value of "high". This result is used with the results of other rules to finally generate the crisp composite output. Obviously, the greater the truth value of "cold", the higher the truth value of "high", though this does not necessarily mean that the output itself will be set to "high" since this is only one rule among many. In some cases, the membership functions can be modified by "hedges" that are equivalent to adverbs. Common hedges include "about", "near", "close to", "approximately", "very", "slightly", "too", "extremely", and "somewhat". These operations may have precise definitions, though the definitions can vary considerably between different implementations. "Very", for one example, squares membership functions; since the membership values are always less than 1, this narrows the membership function. "Extremely" cubes the values to give greater narrowing, while "somewhat" broadens the function by taking the square root.

In practice, the fuzzy rule sets usually have several antecedents that are combined using fuzzy operators, such as AND, OR, and NOT, though again the definitions tend to vary: AND, in one popular definition, simply uses the minimum weight of all the antecedents, while OR uses the maximum value. There is also a NOT operator that subtracts a membership function from 1 to give the "complementary" function.

There are several ways to define the result of a rule, but one of the most common and simplest is the "max-min" inference method, in which the output membership function is given the truth value generated by the premise.

Rules can be solved in parallel in hardware, or sequentially in software. The results of all the rules that have fired are "defuzzified" to a crisp value by one of several methods. There are dozens, in theory, each with various advantages or drawbacks.

The "centroid" method is very popular, in which the "center of mass" of the result provides the crisp value. Another approach is the "height" method, which takes the value of the biggest contributor. The centroid method favors the rule with the output of greatest area, while the height method obviously favors the rule with the greatest output value.

The diagram below demonstrates max-min inferencing and centroid defuzzification for a system with input variables "x", "y", and "z" and an output variable "n". Note that "mu" is standard fuzzy-logic nomenclature for "truth value":

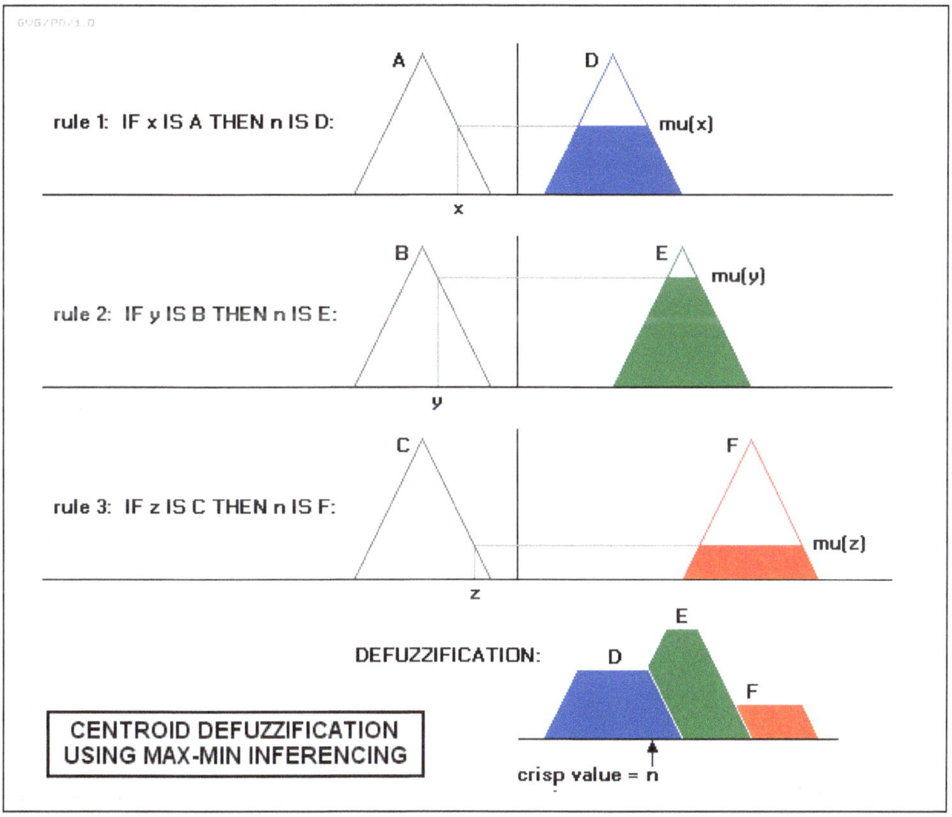

Notice how each rule provides a result as a truth value of a particular membership function for the output variable. In centroid defuzzification the values are OR'd, that is, the maximum value is used and values are not added, and the results are then combined using a centroid calculation.

Fuzzy control system design is based on empirical methods, basically a methodical approach to trial-and-error. The general process is as follows:

- Document the system's operational specifications and inputs and outputs.
- Document the fuzzy sets for the inputs.
- Document the rule set.
- Determine the defuzzification method.
- Run through test suite to validate system, adjust details as required.
- Complete document and release to production.

As a general example, consider the design of a fuzzy controller for a steam turbine. The block diagram of this control system appears as follows:

The input and output variables map into the following fuzzy set:

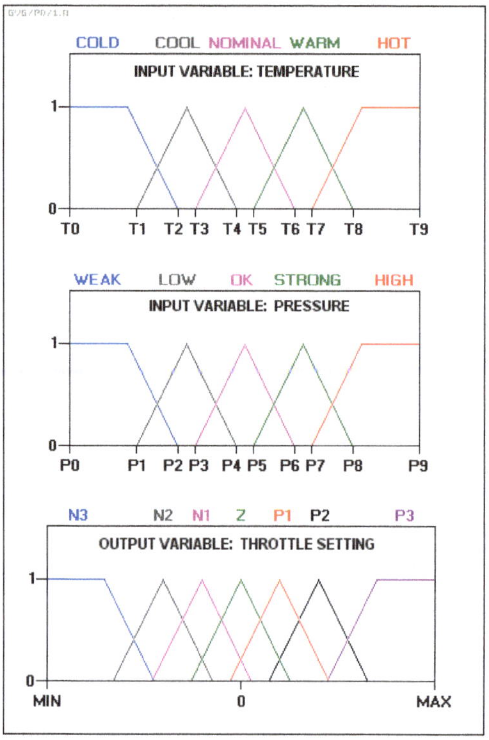

N3: Large negative.

N2: Medium negative.

N1: Small negative.

Z: Zero.

P1: Small positive.

P2: Medium positive.

P3: Large positive.

The rule set includes such rules as:

rule 1: IF temperature IS cool AND pressure IS weak,
THEN throttle is P3.

rule 2: IF temperature IS cool AND pressure IS low,
THEN throttle is P2.

rule 3: IF temperature IS cool AND pressure IS ok,
THEN throttle is Z.

rule 4: IF temperature IS cool AND pressure IS strong,
THEN throttle is N2.

In practice, the controller accepts the inputs and maps them into their membership functions and truth values. These mappings are then fed into the rules. If the rule specifies an AND relationship between the mappings of the two input variables, as the examples above do, the minimum of the two is used as the combined truth value; if an OR is specified, the maximum is used. The appropriate output state is selected and assigned a membership value at the truth level of the premise. The truth values are then defuzzified. For an example, assume the temperature is in the "cool" state, and the pressure is in the "low" and "ok" states. The pressure values ensure that only rules 2 and 3 fire:

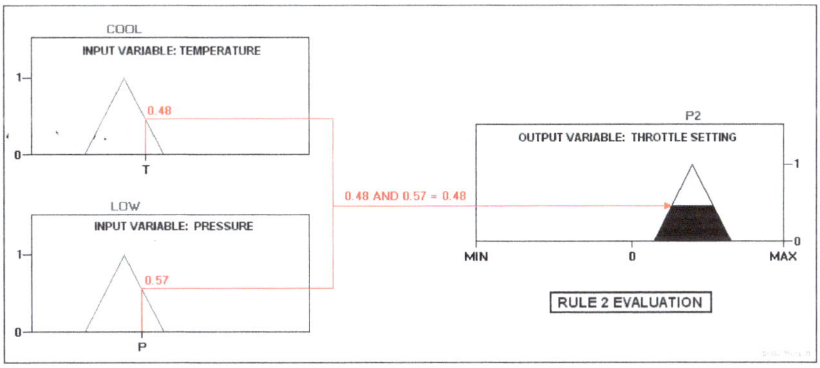

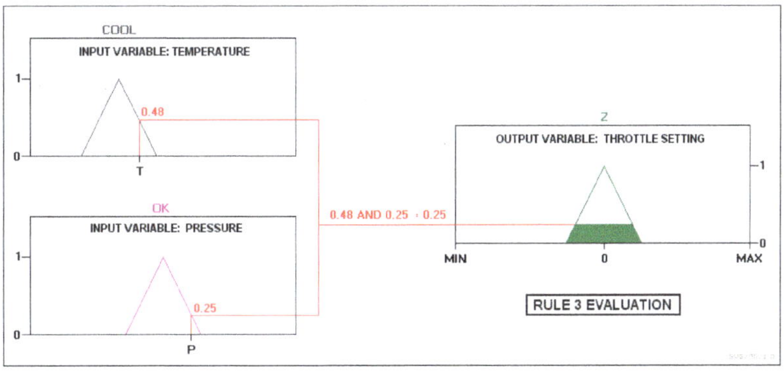

The two outputs are then defuzzified through centroid defuzzification:

```
      |          Z      P2
    1 -+          *       *
      |         * *     * *
      |        *   *   *   *
      |       *     * *     *
      |      *       222222222
      |     *       22222222222
      |    333333332222222222222
      +---333333332222222222222222-->
                     ^

                    +150
```

The output value will adjust the throttle and then the control cycle will begin again to generate the next value .

Building a fuzzy controller

Consider implementing with a microcontroller chip a simple feedback controller:

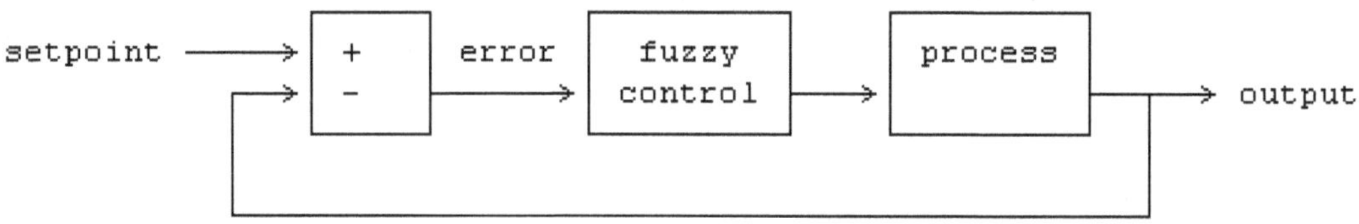

A fuzzy set is defined for the input error variable "e", and the derived change in error, "delta", as well as the "output", as follows:

```
LP:  large positive
SP:  small positive
ZE:  zero
SN:  small negative
LN:  large negative
```

If the error ranges from -1 to +1, with the analog-to-digital converter used having a resolution of 0.25, then the input variable's fuzzy set (which, in this case, also applies to the output variable) can be described very simply as a table, with the error / delta / output values in the top row and the truth values for each membership function arranged in rows beneath:

	-1	-0.75	-0.5	-0.25	0	0.25	0.5	0.75	1
mu(LP)	0	0	0	0	0	0	0.3	0.7	1
mu(SP)	0	0	0	0	0.3	0.7	1	0.7	0.3
mu(ZE)	0	0	0.3	0.7	1	0.7	0.3	0	0
mu(SN)	0.3	0.7	1	0.7	0.3	0	0	0	0
mu(LN)	1	0.7	0.3	0	0	0	0	0	0

—or, in graphical form (where each "X" has a value of 0.1):

```
          LN           SN           ZE           SP           LP
      +-----------------------------------------------------------+
      |                                                           |
-1.0  | XXXXXXXXXX   XXX          :            :            :     |
-0.75 | XXXXXXX      XXXXXXX      :            :            :     |
-0.5  | XXX          XXXXXXXXXX   XXX          :            :     |
-0.25 | :            XXXXXXX      XXXXXXX      :            :     |
0.0   | :            XXX          XXXXXXXXXX   XXX          :     |
0.25  | :            :            XXXXXXX      XXXXXXX      :     |
0.5   | :            :            XXX          XXXXXXXXXX   XXX   |
0.75  | :            :            :            XXXXXXX      XXXXXXX |
1.0   | :            :            :            XXX          XXXXXXXXXX |
      |                                                           |
      +-----------------------------------------------------------+
```

Suppose this fuzzy system has the following rule base:

```
rule 1:  IF e = ZE AND delta = ZE THEN output = ZE
rule 2:  IF e = ZE AND delta = SP THEN output = SN
rule 3:  IF e = SN AND delta = SN THEN output = LP
rule 4:  IF e = LP OR  delta = LP THEN output = LN
```

These rules are typical for control applications in that the antecedents consist of the logical combination of the error and error-delta signals, while the consequent is a control command output. The rule outputs can be defuzzified using a discrete centroid computation:

```
SUM( I = 1 TO 4 OF ( mu(I) * output(I) ) ) / SUM( I = 1 TO 4 OF mu(I) )
```

Now, suppose that at a given time we have:

```
e     = 0.25
delta = 0.5
```

Then this gives:

```
______________________

           e     delta
______________________

mu(LP)     0      0.3
mu(SP)    0.7      1
mu(ZE)    0.7     0.3
mu(SN)     0       0
mu(LN)     0       0
______________________
```

```
rule 1:  IF e = ZE AND delta = ZE THEN output = ZE

   mu(1)     = MIN( 0.7, 0.3 ) = 0.3
   output(1) = 0
```

-- where:

- mu(1): Truth value of the result membership function for rule 1. In terms of a centroid calculation, this is the "mass" of this result for this discrete case.
- output(1): Value (for rule 1) where the result membership function (ZE) is maximum over the output variable fuzzy set range. That is, in terms of a centroid calculation, the location of the "center of mass" for this individual result. This value is independent of the value of "mu". It simply identifies the location of ZE along the output range.

The other rules give:

```
rule 2:  IF e = ZE AND delta = SP THEN output = SN

   mu(2)     = MIN( 0.7, 1 ) = 0.7
   output(2) = -0.5


rule 3: IF e = SN AND delta = SN THEN output = LP

   mu(3)     = MIN( 0.0, 0.0 ) = 0
   output(3) = 1


rule 4: IF e = LP OR  delta = LP THEN output = LN

   mu(4)     = MAX( 0.0, 0.3 ) = 0.3
   output(4) = -1
```

The centroid computation yields:

$$\frac{mu(1).output(1) + mu(2).output(2) + mu(3).output(3) + mu(4).output(4)}{mu(1) + mu(2) + mu(3) + mu(4)}$$
$$= \frac{(0.3 * 0) + (0.7 * -0.5) + (0 * 1) + (0.3 * -1)}{0.3 + 0.7 + 0 + 0.3}$$

$= -0.5$ for the final control output. Simple. Of course the hard part is figuring out what rules actually work correctly in practice.

If you have problems figuring out the centroid equation, remember that a centroid is defined by summing all the moments (location times mass) around the center of gravity and equating the sum to zero. So if So if X0 is the center of gravity, Xi is the location of each mass, and Mi is each mass, this gives:

$$0 = (X_1 - X_0) * M_1 + (X_2 - X_0) * M_2 + \ldots + (X_n - X_0) * M_n$$
$$0 = (X_1 * M_1 + X_2 * M_2 + \ldots + X_n * M_n) - X_0 * (M_1 + M_2 + \ldots + M_n)$$
$$X_0 * (M_1 + M_2 + \ldots + M_n) = X_1 * M_1 + X_2 * M_2 + \ldots + X_n * M_n$$
$$X_0 = \frac{X_1 * M_1 + X_2 * M_2 + \ldots + X_n * M_n}{M_1 + M_2 + \ldots + M_n}$$

In our example, the values of mu correspond to the masses, and the values of X to location of the masses (mu, however, only 'corresponds to the masses' if the initial 'mass' of the output functions are all the same/equivalent. If they are not the same, i.e. some are narrow triangles, while others maybe wide trapizoids or shouldered triangles, then the mass or area of the output function must be known or calculated. It is this mass that is then scaled by mu and multiplied by its location X_i). This system can be implemented on a standard microprocessor, but dedicated fuzzy chips are now available. For example, Adaptive Logic INC of San Jose, California, sells a "fuzzy chip", the AL220, that can accept four analog inputs and generate four analog outputs. A block diagram of the chip is shown below:

```
             +---------+                                +-------+
analog --4-->| analog  |                                | mux / +--4--> analog
  in         |   mux   |                                |  SH   |         out
             +----+----+                                +-------+
                  |                                         ^
                  V                                         |
           +-------------+                               +--+--+
           | ADC / latch |                               | DAC |
           +------+------+                               +-----+
                  |                                         ^
                  |                                         |
                  8         +-------------------------------+
                  |         |                               |
                  |         V                               |
                  |   +-----------+       +-------------+   |
                  +-->| fuzzifier |       | defuzzifier +---+
                      +-----+-----+       +-------------+
                            |                    ^
                            |  +-------------+   |
                            |  |    rule     |   |
                            +->|  processor  +---+
                               | (50 rules)  |
                               +------+------+
                                      |
                               +------+------+
                               |  parameter  |
                               |   memory    |
                               |   256 x 8   |
                               +-------------+

ADC:  analog-to-digital converter
DAC:  digital-to-analog converter
SH:   sample/hold
```

Antilock Brakes

As a first example, consider an anti-lock braking system, directed by a microcontroller chip. The microcontroller has to make decisions based on brake temperature, speed, and other variables in the system.

The variable "temperature" in this system can be subdivided into a range of "states": "cold", "cool", "moderate", "warm", "hot", "very hot". The transition from one state to the next is hard to define.

An arbitrary static threshold might be set to divide "warm" from "hot". For example, at exactly 90 degrees, warm ends and hot begins. But this would result in a discontinuous change when the input value passed over that threshold. The transition wouldn't be smooth, as would be required in braking situations.

The way around this is to make the states *fuzzy*. That is, allow them to change gradually from one state to the next. In order to do this there must be a dynamic relationship established between different factors.

We start by defining the input temperature states using "membership functions":

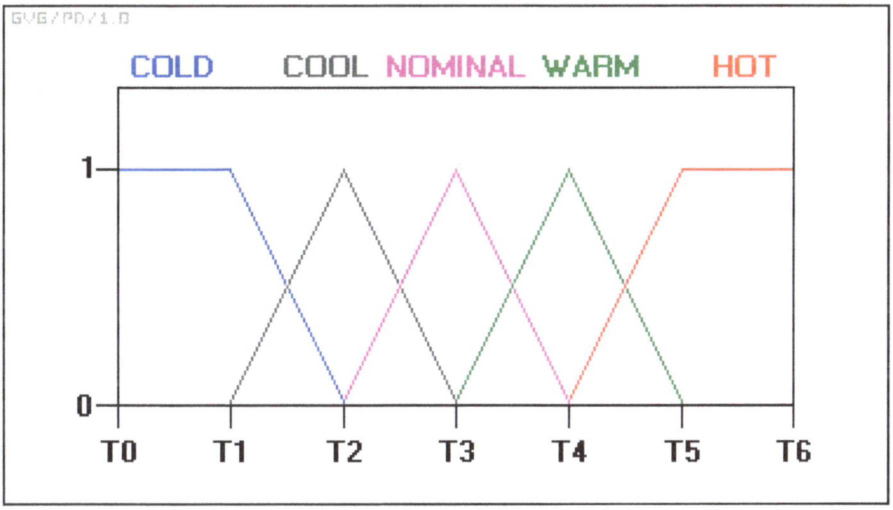

With this scheme, the input variable's state no longer jumps abruptly from one state to the next. Instead, as the temperature changes, it loses value in one membership function while gaining value in the next. In other words, its ranking in the category of cold decreases as it becomes more highly ranked in the warmer category.

At any sampled timeframe, the "truth value" of the brake temperature will almost always be in some degree part of two membership functions: i.e.: '0.6 nominal and 0.4 warm', or '0.7 nominal and 0.3 cool', and so on.

The above example demonstrates a simple application, using the abstraction of values from multiple values. This only represents one kind of data, however, in this case, temperature.

Adding additional sophistication to this braking system, could be done by additional factors such as traction, speed, inertia, set up in dynamic functions, according to the designed fuzzy system.

Logical Interpretation of Fuzzy Control

In spite of the appearance there are several difficulties to give a rigorous logical interpretation of the *IF-THEN* rules. As an example, interpret a rule as *IF (temperature is "cold") THEN (heater is "high")* by the first order formula *Cold(x) High(y)* and assume that r is an input such that *Cold(r)* is false. Then the formula *Cold(r) High(t)* is true for any *t* and therefore any *t* gives a correct control given *r*. A rigorous logical justification of fuzzy control is given in Hájek's book (see Chapter 7) where fuzzy control is represented as a theory of Hájek's basic logic.Also in Gerla 2005 another logical approach to fuzzy control is proposed based on fuzzy logic programming.Indeed, denote by *f* the fuzzy function arising of an IF-THEN systems of rules. Then we can translate this system into a fuzzy program P containing a series of rules whose head is "Good(x,y)". The interpretation of this predicate in the least fuzzy Herbrand model of P coincides with f. This gives further useful tools to fuzzy control.

FUZZY FINITE ELEMENT

The fuzzy finite element method combines the well-established finite element method with the concept of fuzzy numbers, the latter being a special case of a fuzzy set. The advantage of using fuzzy numbers instead of real numbers lies in the incorporation of uncertainty (on material properties, parameters, geometry, initial conditions, etc.) in the finite element analysis.

One way to establish a fuzzy finite element (FE) analysis is to use existing FE software (in-house or commercial) as an inner-level module to compute a deterministic result, and to add an outer-level loop to handle the fuzziness (uncertainty). This outer-level loop comes down to solving an optimization problem. If the inner-level deterministic module produces monotonic behavior with respect to the input variables, then the outer-level optimization problem is greatly simplified, since in this case the extrema will be located at the vertices of the domain.

Parabolic Equation

Traditional finite element method is a well-established method to solve various problems of science and engineering. Different authors have used various methods to solve governing differential equation of heat conduction problem. In this study, heat conduction in a circular rod has been considered which is made up of two different materials viz. aluminum and copper. In earlier studies parameters in the differential equation have been taken as fixed numbers which actually may not. Those parameters are found in general by some measurements or experiments. So the material properties are actually uncertain and may be considered to vary in an interval or as fuzzy and in that case complex interval arithmetic or fuzzy arithmetic has to be considered in the analysis. Hence interval/fuzzy arithmetic is applied in the finite element method to solve a steady state heat conduction problem. Application of fuzzy finite element method in the said problem gives fuzzy system of linear equations in general. Here we have also proposed new methods to handle such type of fuzzy system of linear equations. Corresponding results are computed and has been reported here.

REFERENCES

1. H. Bustince, "Indicator of inclusion grade for interval-valued fuzzy sets: Application to approximate reasoning based on interval-valued fuzzy sets," *Int'l. J. of Approximate Reasoning*, vol. 23, pp. 137–209, 2000.
2. O. Castillo and P. Melin, *Type-2 Fuzzy Logic Theory and Applications*, Springer-Verlag, Berlin, 2008.
3. M. B. Gorzalczany, "A Method of Inference in Approximate Reasoning Based on Interval-Valued Fuzzy Sets," *Fuzzy Sets and Systems*, vol. 21, pp. 1–17, 1987.
4. H. Hagras, "Type-2 FLCs: A new generation of fuzzy controllers," *IEEE Computational Intelligence Magazine*, vol. 2, pp. 30–43, February 2007.
5. R. John and S. Coupland, "Type-2 fuzzy logic: a historical view," *IEEE Computational Intelligence Magazine*, vol. 2, pp. 57–62, February 2007.
6. N. N. Karnik and J. M. Mendel, "Operations on Type-2 Fuzzy Sets," *Fuzzy Sets and Systems*, vol. 122, pp. 327–348, 2001.
7. N. N. Karnik and J. M. Mendel, "Centroid of a type-2 fuzzy set," *Information Sciences*, vol. 132, pp. 195–220, 2001.
8. Q. Liang and J. M. Mendel, "Interval Type-2 Fuzzy Logic Systems: Theory and Design," *IEEE Trans. on Fuzzy Systems*, vol. 8, pp. 535–550, 2000.
9. F. Liu and J. M. Mendel, "Aggregation Using the Fuzzy Weighted Average, as Computed by the KM Algorithms," *IEEE Trans. on Fuzzy Systems*, vol. 16, pp. 1–12, February 2008.
10. F. Liu and J. M. Mendel, "Encoding words into interval type-2 fuzzy sets using an interval approach," *IEEE Trans. on Fuzzy Systems*, vol. 16, pp. 1503–1521, December 2008.
11. M. Melgarejo, "Implementing interval type-2 fuzzy processors," *IEEE Computational Intelligence Magazine*, vol. 2, pp. 63–71, February 2007.
12. J. M. Mendel, *Uncertain Rule-Based Fuzzy Logic Systems: Introduction and New Directions*, Prentice-Hall, Upper-Saddle River, NJ, 2001.
13. J. M. Mendel, "Fuzzy Sets for Words: a New Beginning," *Proc. IEEE FUZZ Conference*, St. Louis, MO, May 26–28, 2003, pp. 37–42.
14. J. M. Mendel, "Type-2 fuzzy sets and systems: an overview," *IEEE Computational Intelligence Magazine*, vol. 2, pp. 20–29, February 2007.
15. J. M. Mendel, "Advances in type-2 fuzzy sets and systems," *Information Sciences*, Vol. 177, pp. 84–110, 2007.
16. J. M. Mendel, "Computing with words: Zadeh, Turing, Popper and Occam," *IEEE Computational Intelligence Magazine*, vol. 2, pp. 10–17, November 2007.
17. J. M. Mendel and R. I. John, """Type-2 Fuzzy Sets Made Simple," *IEEE Trans. on*

Fuzzy Systems, vol. 10, pp. 117–127, April 2002.

18. J. M. Mendel, R. I. John and F. Liu, "Interval type-2 fuzzy logic systems made simple," *IEEE Trans. on Fuzzy Systems*, vol. 14, pp. 808–821, December 2006.
19. K. Popper, *The Logic of Scientific Discovery* (translation of Logik der Forschung), Hutchinson, London, 1959.
20. F. Rhee, "Uncertain fuzzy clustering: Insights and recommendations," *IEEE Computational Intelligence Magazine*, vol. 2, pp. 44–56, February 2007.
21. J. T. Rickard, J. Aisbett, G. Gibbon and D. Morgenthaler, "Fuzzy subsethood for type-n fuzzy sets," *NAFIPS 2008*, Paper # 60101, New York City, May 2008.
22. D. Wu and J. M. Mendel, "Uncertainty measures for interval type-2 fuzzy sets," *Information Sciences*, vol. 177, pp. 5378–5393, 2007.
23. D. Wu and J. M. Mendel, "Aggregation Using the Linguistic Weighted Average and Interval Type-2 Fuzzy Sets," *IEEE Trans. on Fuzzy Systems*, vol. 15, pp. 1145–1161, December 2007.
24. D. Wu and J. M. Mendel, "A Vector Similarity Measure for Interval Type-2 Fuzzy Sets and Type-1 Fuzzy Sets," *Information Sciences*, vol. 178, pp. 381–402, 2008.
25. D. Wu and J. M. Mendel, "A comparative study of ranking methods, similarity measures and uncertainty measures for interval type-2 fuzzy sets," *Information Sciences*, to appear in 2009.
26. H. Wu and J. M. Mendel, "Uncertainty Bounds and Their Use in the Design of Interval Type-2 Fuzzy Logic Systems," *IEEE Trans. on Fuzzy Systems*, vol. 10, pp. 622–639, Oct. 2002.
27. L. A. Zadeh, "The Concept of a Linguistic Variable and Its Application to Approximate Reasoning–1," *Information Sciences*, vol. 8, pp. 199–249, 1975.
28. L. A. Zadeh, "Fuzzy logic = computing with words," *IEEE Trans. on Fuzzy Systems*, vol. 4, pp. 103–111, 1996.
29. L. A. Zadeh, "From computing with numbers to computing with words—from manipulation of measurements to manipulation of perceptions," *IEEE Trans. on Circuits and Systems–1, Fundamental Theory and Applications*, vol. 4, pp. 105–119, 1999.
30. L. A. Zadeh, "Toward human level machine intelligence—is it achievable? The need for a new paradigm shift," *IEEE Computational Intelligence Magazine*, vol. 3, pp. 11–22, August 2008.
31. S. -M. Zhou, J. M. Garibaldi, R. I. John and F. Chiclana, "On constructing parsimonious type-2 fuzzy logic systems via influential rule selection," *IEEE Trans. on Fuzzy Systems*, vol.17, no.3, pp. 654–667, 2009.
32. O. Salazar and J. Soriano, "Generating embedded type-1 fuzzy sets by means of convex combination," in Proceedings of the 2013 IFSA World Congress NAFIPS Annual Meeting, Edmonton, Canada, Jun. 2013, pp. 51–56.
33. O. Salazar, J. Soriano, and H. Serrano, "A short note on the centroid of an interval type-2 fuzzy set," in Proceedings of IEEE 2012 Workshop on Engineering Applications (WEA), Bogota, Colombia, May 2012, pp. 1–4
34. O. Salazar, and J. Soriano, "Convex combination and its application to fuzzy sets and interval-valued fuzzy sets I," Applied Mathematical Sciences, vol. 9, no. 22, pp. 1061–1068, 2015

35. O. Salazar, and J. Soriano, "Convex combination and its application to fuzzy sets and interval-valued fuzzy sets II," Applied Mathematical Sciences, vol. 9, no. 22, pp. 1069–1076, 2015

36. Alkhazaleh, S. and Salleh, A.R. Fuzzy Soft Multiset Theory, Abstract and Applied Analysis, 2012, article ID 350600, 20 p.

37. Alkhazaleh, S., Salleh, A.R. and Hassan, N. Soft Multisets Theory, Applied Mathematical Sciences, v. 5, No. 72, 2011, pp. 3561–3573

38. Atanassov, K. T. (1983) Intuitionistic fuzzy sets, VII ITKR's Session, Sofia (deposited in Central Sci.-Technical Library of Bulg. Acad. of Sci., 1697/84) (in Bulgarian)

39. Atanasov, K. (1986) Intuitionistic Fuzzy Sets, Fuzzy Sets and Systems, v. 20, No. 1, pp. 87–96

*40. Bezdek, J.C. (1978). "Fuzzy partitions and relations and axiomatic basis for clustering". Fuzzy Sets and Systems. **1**. pp. 111–127.*

41. Blizard, W.D. (1989) Real-valued Multisets and Fuzzy Sets, Fuzzy Sets and Systems, v. 33, pp. 77–97

42. Brown, J.G. (1971) A Note on Fuzzy Sets, Information and Control, v. 18, pp. 32–39

43. Burgin, M. Theory of Named Sets as a Foundational Basis for Mathematics, in Structures in Mathematical Theories, San Sebastian, 1990, pp. 417-420

44. Burgin M. and Chunihin, A. (1997) Named Sets in the Analysis of Uncertainty, in Methodological and Theoretical Problems of Mathematics and Information Sciences, Kiev, pp. 72-85

45. Chapin, E.W. (1974) Set-valued Set Theory, I, Notre Dame J. Formal Logic, v. 15, pp. 619–634

46. Chapin, E.W. (1975) Set-valued Set Theory, II, Notre Dame J. Formal Logic, v. 16, pp. 255–267

47. Chris Cornelis, Martine De Cock and Etienne E. Kerre, Intuitionistic fuzzy rough sets: at the crossroads of imperfect knowledge, Expert Systems, v. 20, issue 5, pp. 260–270, 2003

48. Cornelis, C., Deschrijver, C., and Kerre, E. E. (2004) Implication in intuitionistic and interval-valued fuzzy set theory: construction, classification, application, International Journal of Approximate Reasoning, v. 35, pp. 55–95

49. De Cock, Martine; Bodenhofer, Ulrich; Kerre, Etienne E. (1–4 October 2000). Modelling Linguistic Expressions Using Fuzzy Relations. Proceedings of the 6th International Conference on Soft Computing. Iizuka, Japan. pp. 353–360.

50. Demirci, M. (1999) Genuine Sets, Fuzzy Sets and Systems, v. 105, pp. 377–384

*51. Deschrijver, G.; Kerre, E.E. (2003). "On the relationship between some extensions of fuzzy set theory". Fuzzy Sets and Systems. **133** (2): 227–235.*

52. Didier Dubois, Henri M. Prade, ed. (2000). Fundamentals of fuzzy sets. The Handbooks of Fuzzy Sets Series. 7. Springer. ISBN 978-0-7923-7732-0.

53. Feng F. Generalized Rough Fuzzy Sets Based on Soft Sets, Soft Computing, July 2010, Volume 14, Issue 9, pp 899–911

54. Gentilhomme, Y. (1968) Les ensembles flous en linguistique, Cahiers Linguistique Theoretique Appliqee, 5, pp. 47–63

55. Gogen, J.A. (1967) L-fuzzy Sets, Journal Math. Analysis Appl., v. 18, pp. 145–174

56. *Gottwald, S. (2006). "Universes of Fuzzy Sets and Axiomatizations of Fuzzy Set Theory. Part I: Model-Based and Axiomatic Approaches". Studia Logica.* ***82*** *(2): 211–244. doi:10.1007/s11225-006-7197-8.. Gottwald, S. (2006). "Universes of Fuzzy Sets and Axiomatizations of Fuzzy Set Theory. Part II: Category Theoretic Approaches". Studia Logica.* ***84****: 23–50. doi:10.1007/s11225-006-9001-1.* preprint..

57. Grattan-Guinness, I. (1975) Fuzzy membership mapped onto interval and many-valued quantities. Z. Math. Logik. Grundladen Math. 22, pp. 149–160.

58. Grzymala-Busse, J. Learning from examples based on rough multisets, in Proceedings of the 2nd International Symposium on Methodologies for Intelligent Systems, Charlotte, NC, USA, 1987, pp. 325–332

59. Gylys, R. P. (1994) Quantal sets and sheaves over quantales, Liet. Matem. Rink., v. 34, No. 1, pp. 9–31.

60. *Ulrich Höhle, Stephen Ernest Rodabaugh, ed. (1999). Mathematics of fuzzy sets: logic, topology, and measure theory. The Handbooks of Fuzzy Sets Series.* ***3****. Springer. ISBN 978-0-7923-8388-8.*

61. Jahn, K. U. (1975) Intervall-wertige Mengen, Math.Nach. 68, pp. 115–132

62. *Kerre, E.E. (2001). B. Reusch; K-H. Temme, eds. "A first view on the alternatives of fuzzy set theory". Computational Intelligence in Theory and Practice. Heidelberg: Physica-Verlag: 55–72. ISBN 3-7908-1357-5.*

63. *George J. Klir; Bo Yuan (1995). Fuzzy sets and fuzzy logic: theory and applications. Prentice Hall. ISBN 978-0-13-101171-7.*

64. *Kuzmin, V.B. (1982). "Building Group Decisions in Spaces of Strict and Fuzzy Binary Relations" (in Russian). Nauka, Moscow.*

65. Lake, J. (1976) Sets, fuzzy sets, multisets and functions, J. London Math. Soc., II Ser., v. 12, pp. 323–326

66. Meng, D., Zhang, X. and Qin, K. Soft rough fuzzy sets and soft fuzzy rough sets, 'Computers & Mathematics with Applications', v. 62, issue 12, 2011, pp. 4635–4645

67. Miyamoto, S. Fuzzy Multisets and their Generalizations, in 'Multiset Processing', LNCS 2235, pp. 225–235, 2001

68. Molodtsov, O. (1999) Soft set theory – first results, Computers & Mathematics with Applications, v. 37, No. 4/5, pp. 19–31

69. Moore, R.E. Interval Analysis, New York, Prentice-Hall, 1966

70. Nakamura, A. (1988) Fuzzy rough sets, 'Notes on Multiple-valued Logic in Japan', v. 9, pp. 1–8

71. Narinyani, A.S. Underdetermined Sets – A new datatype for knowledge representation, Preprint 232, Project VOSTOK, issue 4, Novosibirsk, Computing Center, USSR Academy of Sciences, 1980

72. Pedrycz, W. Shadowed sets: representing and processing fuzzy sets, IEEE Transactions on System, Man, and Cybernetics, Part B, 28, 103–109, 1998.

73. Radecki, T. Level Fuzzy Sets, 'Journal of Cybernetics', Volume 7, Issue 3-4, 1977

74. Radzikowska, A.M. and Etienne E. Kerre, E.E. On L-Fuzzy Rough Sets, Artificial Intelligence and Soft Computing – ICAISC 2004, 7th International Conference, Zakopane, Poland, June 7–11, 2004, Proceedings; 01/2004

75. *Salii, V.N. (1965). "Binary L-relations". Izv. Vysh. Uchebn. Zaved. Matematika (in Russian). **44** (1): 133–145.*

76. Sambuc, R. Fonctions φ-floues: Application a l"aide au diagnostic en pathologie thyroidienne, Ph. D. Thesis Univ. Marseille, France, 1975.

77. Seising, Rudolf: The Fuzzification of Systems. The Genesis of Fuzzy Set Theory and Its Initial Applications—Developments up to the 1970s (Studies in Fuzziness and Soft Computing, Vol. 216) Berlin, New York, [et al.]: Springer 2007.

78. Smith, N.J.J. (2004) Vagueness and blurry sets, ‹J. of Phil. Logic", 33, pp. 165–235

79. Thomas, K.V. and L. S. Nair, Rough intuitionistic fuzzy sets in a lattice, ‹International Mathematical Forum", Vol. 6, 2011, no. 27, 1327–1335

80. Yager, R. R. (1986) On the Theory of Bags, International Journal of General Systems, v. 13, pp. 23–37

81. Yao, Y.Y., Combination of rough and fuzzy sets based on α-level sets, in: Rough Sets and Data Mining: Analysis for Imprecise Data, Lin, T.Y. and Cercone, N. (Eds.), Kluwer Academic Publishers, Boston, pp. 301–321, 1997.

82. Y. Y. Yao, A comparative study of fuzzy sets and rough sets, Information Sciences, v. 109, Issue 1-4, 1998, pp. 227 – 242

83. Zadeh, L. (1975) The concept of a linguistic variable and its application to approximate reasoning–I, Inform. Sci., v. 8, pp. 199–249

84. *Hans-Jürgen Zimmermann (2001). Fuzzy set theory—and its applications (4th ed.). Kluwer. ISBN 978-0-7923-7435-0.*

85. Gianpiero Cattaneo and Davide Ciucci, "Heyting Wajsberg Algebras as an Abstract Environment Linking Fuzzy and Rough Sets" in J.J. Alpigini et al. (Eds.): RSCTC 2002, LNAI 2475, pp. 77–84, 2002. doi:10.1007/3-540-45813-1_10

CHAPTER 4

Classical Logic

Chapter Contents

- **Introduction**
- **Double Negation**
- **Law Of Noncontradiction**
- **Monotonicity Of Entailment**
- **Commutativity Of Conjunction**
- **De Morgan's Laws**
- **Sorites Paradox**

A classification of formal logic, characterized by the non-contradiction of variables and the elimination of double-negative variables; the recognition of multiple paths to the same answer; that additional assumptions may be made to expand upon the original argument, and that any logical argument may be interchanged freely with another one as long as they are all both true statements.

INTRODUCTION

Classical logic (or standard logic) is an intensively studied and widely used class of formal logics. Each logical system in this class shares characteristic properties:

- Law of excluded middle and double negative elimination
- Law of noncontradiction, and the principle of explosion

- Monotonicity of entailment and idempotency of entailment
- Commutativity of conjunction
- De Morgan duality: every logical operator is dual to another

While not entailed by the preceding conditions, contemporary discussions of classical logic normally only include propositional and first-orderlogics.

Classical logic was originally devised as a two-level (bivalent) logical system, with simple semantics for the levels representing "true" and "false".

Examples of Classical Logics

- Aristotle's Organon introduces his theory of syllogisms, which is a logic with a restricted form of judgments: assertions take one of four forms, All Ps are Q, Some Ps are Q, No Ps are Q, and Some Ps are not Q. These judgments find themselves if two pairs of two dual operators, and each operator is the negation of another, relationships that Aristotle summarised with his square of oppositions. Aristotle explicitly formulated the law of the excluded middle and law of non-contradiction in justifying his system, although these laws cannot be expressed as judgments within the syllogistic framework.
- George Boole's algebraic reformulation of logic, his system of Boolean logic;
- The first-order logic found in Gottlob Frege's Begriffsschrift.

Generalized Semantics

With the advent of algebraic logic it became apparent however that classical propositional calculus admits other semantics. In Boolean-valued semantics (for classical propositional logic), the truth values are the elements of an arbitrary Boolean algebra; "true" corresponds to the maximal element of the algebra, and "false" corresponds to the minimal element. Intermediate elements of the algebra correspond to truth values other than "true" and "false". The principle of bivalence holds only when the Boolean algebra is taken to be the two-element algebra, which has no intermediate elements.

Non-Classical Logics

- Computability logic is a semantically constructed formal theory of computability—as opposed to classical logic, which is a formal theory of truth—integrates and extends classical, linear and intuitionistic logics.
- Many-valued logic, including fuzzy logic, which rejects the law of the excluded middle and allows as a truth value any real number between 0 and 1.
- Intuitionistic logic rejects the law of the excluded middle, double negative elimination, and the De Morgan's laws;
- Linear logic rejects idempotency of entailment as well;
- Modal logic extends classical logic with non-truth-functional ("modal") operators.

- Paraconsistent logic (e.g., dialetheism and relevance logic) rejects the law of noncontradiction;
- Relevance logic, linear logic, and non-monotonic logic reject monotonicity of entailment;

In Deviant Logic, Fuzzy Logic: Beyond the Formalism, Susan Haack divided non-classical logics into deviant, quasi-deviant, and extended logics.

Law of excluded middle

In logic, the law of excluded middle (or the principle of excluded middle) is the third of the three classic laws of thought. It states that for any proposition, either that proposition is true, or its negation is true.

The law is also known as the law (or principle) of the excluded third, in Latin principium tertii exclusi. Another Latin designation for this law is tertium non datur: "no third (possibility) is given".

The earliest known formulation is Aristotle's principle of non-contradiction, first proposed in On Interpretation, where he says that of two contradictory propositions (i.e. where one proposition is the negation of the other) one must be true, and the other false. He also states it as a principle in the Metaphysics book 3, saying that it is necessary in every case to affirm or deny, and that it is impossible that there should be anything between the two parts of a contradiction. The principle was stated as a theorem of propositional logic by Russell and Whitehead in Principia Mathematica as:

$*2\cdot 11.\ \vdash .\ p \vee \sim p.$

The principle should not be confused with the semantical principle of bivalence, which states that every proposition is either true or false.

Classic Laws of Thought

The principle of excluded middle, along with its complement, the law of contradiction (the second of the three classic laws of thought), are correlates of the law of identity (the first of these laws).

Analogous Laws

Some systems of logic have different but analogous laws. For some finite n-valued logics, there is an analogous law called the **law of excluded *n*+1th**. If negation is cyclic and "∨" is a "max operator", then the law can be expressed in the object language by (P ∨ ~P ∨ ~~P ∨ ... ∨ ~...~P), where "~...~" represents n−1 negation signs and "∨ ... ∨" n−1 disjunction signs. It is easy to check that the sentence must receive at least one of the n truth values (and not a value that is not one of the n).

Other systems reject the law entirely.

Examples

For example, if P is the proposition:

Socrates is mortal.

then the law of excluded middle holds that the logical disjunction:

Either Socrates is mortal, or it is not the case that Socrates is mortal.

is true by virtue of its form alone. That is, the "middle" position, that Socrates is neither mortal nor not-mortal, is excluded by logic, and therefore either the first possibility (Socrates is mortal) or its negation (it is not the case that Socrates is mortal) must be true.

An example of an argument that depends on the law of excluded middle follows. We seek to prove that there exist two irrational numbers a and b such that

a^b is rational.

It is known that $\sqrt{2}$ is irrational (see proof). Consider the number

$$\sqrt{2}^{\sqrt{2}}.$$

Clearly (excluded middle) this number is either rational or irrational. If it is rational, the proof is complete, and

$$a = \sqrt{2} \text{ and } b = \sqrt{2}.$$

But if $\sqrt{2}^{\sqrt{2}}$ is irrational, then let

$$a = \sqrt{2}^{\sqrt{2}} \text{ and } b = \sqrt{2}.$$

Then

$$a^b = \left(\sqrt{2}^{\sqrt{2}}\right)^{\sqrt{2}} = \sqrt{2}^{(\sqrt{2}\cdot\sqrt{2})} = \sqrt{2}^2 = 2,$$

and 2 is certainly rational. This concludes the proof.

In the above argument, the assertion "this number is either rational or irrational" invokes the law of excluded middle. An intuitionist, for example, would not accept this argument without further support for that statement. This might come in the form of a proof that the number in question is in fact irrational (or rational, as the case may be); or a finite algorithm that could determine whether the number is rational.

The Law in non-Constructive Proofs over the Infinite

The above proof is an example of a non-constructive proof disallowed by intuitionists:

The proof is non-constructive because it doesn"t give specific numbers a and b that satisfy the theorem but only two separate possibilities, one of which must work. (Actually $a = \sqrt{2}^{\sqrt{2}}$ is irrational but there is no known easy proof of that fact.) (Davis 2000:220)

By non-constructive Davis means that "a proof that there actually are mathematic entities satisfying certain conditions would have to provide a method to exhibit explicitly the entities in question." (p. 85). Such proofs presume the existence of a totality that is complete, a notion disallowed by intuitionists when extended to the infinite—for them the infinite can never be completed:

In classical mathematics there occur non-constructive or indirect existence proofs, which intuitionists do not accept. For example, to prove there exists an n such that P(n), the classical mathematician may deduce a contradiction from the assumption for all n, not P(n). Under both the classical and the intuitionistic logic, by reductio ad absurdum this gives not for all n, not P(n). The classical logic allows this result to be transformed into there exists an n such that P(n), but not in general the intuitionistic... the classical meaning, that somewhere in the completed infinite totality of the natural numbers there occurs an n such that P(n), is not available to him, since he does not conceive the natural numbers as a completed totality. (Kleene 1952:49–50)

Indeed, David Hilbert and Luitzen E. J. Brouwer both give examples of the law of excluded middle extended to the infinite. Hilbert's example: "the assertion that either there are only finitely many prime numbers or there are infinitely many" (quoted in Davis 2000:97); and Brouwer"s: "Every mathematical species is either finite or infinite." (Brouwer 1923 in van Heijenoort 1967:336).

In general, intuitionists allow the use of the law of excluded middle when it is confined to discourse over finite collections (sets), but not when it is used in discourse over infinite sets (e.g. the natural numbers). Thus intuitionists absolutely disallow the blanket assertion: "For all propositions P concerning infinite sets D: P or ~P" (Kleene 1952:48).

Putative counterexamples to the law of excluded middle include the liar paradox or Quine"s Paradox. Certain resolutions of these paradoxes, particularly Graham Priest"s dialetheism as formalised in LP, have the law of excluded middle as a theorem, but resolve out the Liar as both true and false. In this way, the law of excluded middle is true, but because truth itself, and therefore disjunction, is not exclusive, it says next to nothing if one of the disjuncts is paradoxical, or both true and false.

Aristotle

Aristotle wrote that ambiguity can arise from the use of ambiguous names, but cannot exist in the facts themselves:

It is impossible, then, that "being a man" should mean precisely "not being a man", if "man" not only signifies something about one subject but also has one significance. ... And it will not be possible to be and not to be the same thing, except in virtue of an ambiguity, just as if one whom we call "man", and others were to call "not-man"; but the point in question is not this, whether the same thing can at the same time be and not be a man in name, but whether it can be in fact. (Metaphysics 4.4, W.D. Ross (trans.), GBWW 8, 525–526).

Aristotle"s assertion that "...it will not be possible to be and not to be the same thing", which would be written in propositional logic as $\neg (P \wedge \neg P)$, is a statement modern logicians could call the law of excluded middle $(P \vee \neg P)$, as distribution of the negation of Aristotle's assertion makes them equivalent, regardless that the former claims that no statement is both true and false, while the latter requires that any statement is either true or false.

However, Aristotle also writes, "since it is impossible that contradictories should be at the same time true of the same thing, obviously contraries also cannot belong at the same time to the same thing" (Book IV, CH 6, p. 531). He then proposes that "there cannot be an intermediate between contradictories, but of one subject we must either affirm or deny any one predicate" (Book IV, CH 7, p. 531). In the context of Aristotle"s traditional logic, this is a remarkably precise statement of the law of excluded middle, $P \vee \neg P$.

Leibniz

Its usual form, "Every judgment is either true or false" [footnote 9]..."(from Kolmogorov in van Heijenoort, p. 421) footnote 9: "This is Leibniz's very simple formulation (see Nouveaux Essais, IV,2)...." (ibid p 421)

Bertrand Russell and Principia Mathematica

Bertrand Russell asserts a distinction between the "law of excluded middle" and the "law of noncontradiction". In The Problems of Philosophy, he cites three "Laws of Thought" as more or less "self-evident" or "a priori" in the sense of Aristotle:

1. Law of identity: "Whatever is, is."
2. Law of noncontradiction: "Nothing can both be and not be."
3. Law of excluded middle: "Everything must either be or not be."

These three laws are samples of self-evident logical principles... (p. 72)

It is correct, at least for bivalent logic—i.e. it can be seen with a Karnaugh map—that Russell's Law (2) removes "the middle" of the inclusive-or used in his law (3). And this is the point of Reichenbach"s demonstration that some believe the exclusive-or should take the place of the inclusive-or.

About this issue (in admittedly very technical terms) Reichenbach observes:

The tertium non datur

29. $(x)[f(x) \vee \sim f(x)]$

is not exhaustive in its major terms and is therefore an inflated formula. This fact may perhaps explain why some people consider it unreasonable to write (29) with the inclusive-'or', and want to have it written with the sign of the exclusive-'or'

30. $(x)[f(x) \oplus \sim f(x)]$, where the symbol "$\oplus$" signifies exclusive-or

in which form it would be fully exhaustive and therefore nomological in the narrower sense. (Reichenbach, p. 376)

In line (30) the "(x)" means "for all" or "for every", a form used by Russell and Reichenbach; today the symbolism is usually $\forall x$. Thus an example of the expression would look like this:

- (pig): (Flies(pig) $\oplus$ ~Flies(pig))

- (For all instances of "pig" seen and unseen): ("Pig does fly" or "Pig does not fly" but not both simultaneously)

A formal definition from Principia Mathematica

Principia Mathematica (PM) defines the law of excluded middle formally:

*2.1 : ~p ∨ p (PM p. 101)

Example: Either it is true that "this is red", or it is true that "this is not red". Hence it is true that "this is red or this is not red". (See below for more about how this is derived from the primitive axioms).

So just what is "truth" and "falsehood"? At the opening PM quickly announces some definitions:

Truth-values. The "truth-values" of a proposition is truth if it is true and falsehood if it is false* [*This phrase is due to Frege]...the truth-value of "p ∨ q" is truth if the truth-value of either p or q is truth, and is falsehood otherwise ... that of "~ p" is the opposite of that of p..." (p. 7-8)

This is not much help. But later, in a much deeper discussion, ("Definition and systematic ambiguity of Truth and Falsehood" Chapter II part III, p. 41 ff) PM defines truth and falsehood in terms of a relationship between the "a" and the "b" and the "percipient". For example "This 'a' is 'b'" (e.g. "This 'object a' is 'red'") really means "'object a' is a sense-datum" and "'red' is a sense-datum", and they "stand in relation" to one another and in relation to "I". Thus what we really mean is: "I perceive that 'This object a is red'" and this is an undeniable-by-3rd-party "truth".

PM further defines a distinction between a "sense-datum" and a "sensation":

That is, when we judge (say) "this is red", what occurs is a relation of three terms, the mind, and "this", and "red". On the other hand, when we perceive "the redness of this", there is a relation of two terms, namely the mind and the complex object "the redness of this" (pp. 43–44).

Russell reiterated his distinction between "sense-datum" and "sensation" in his book The Problems of Philosophy (1912) published at the same time as PM (1910–1913):

Let us give the name of "sense-data" to the things that are immediately known in sensation: such things as colours, sounds, smells, hardnesses, roughnesses, and so on. We shall give the name "sensation" to the experience of being immediately aware of these things... The colour itself is a sense-datum, not a sensation. (p. 12)

Russell further described his reasoning behind his definitions of "truth" and "falsehood" in the same book (Chapter XII Truth and Falsehood).

Consequences of the law of excluded middle in Principia Mathematica

From the law of excluded middle, formula ✱2.1 in Principia Mathematica, Whitehead and Russell derive some of the most powerful tools in the logician's argumentation toolkit. (In Principia Mathematica, formulas and propositions are identified by a leading asterisk and two numbers, such as "✱2.1".)

✱2.1 ~p ∨ p "This is the Law of excluded middle" (PM, p. 101).

The proof of ✱2.1 is roughly as follows: "primitive idea" 1.08 defines p → q = ~p ∨ q. Substituting p for q in this rule yields p → p = ~p ∨ p. Since p → p is true (this is Theorem 2.08, which is proved separately), then ~p ∨ p must be true.

✱2.11 p ∨ ~p (Permutation of the assertions is allowed by axiom 1.4)

✱2.12 p → ~(~p) (Principle of double negation, part 1: if "this rose is red" is true then it's not true that "'this rose is not-red' is true".)

✱2.13 p ∨ ~{~(~p)} (Lemma together with 2.12 used to derive 2.14)

✱2.14 ~(~p) → p (Principle of double negation, part 2)

✱2.15 (~p → q) → (~q → p) (One of the four "Principles of transposition". Similar to 1.03, 1.16 and 1.17. A very long demonstration was required here.)

✱2.16 (p → q) → (~q → ~p) (If it's true that "If this rose is red then this pig flies" then it's true that "If this pig doesn't fly then this rose isn't red.")

✱2.17 (~p → ~q) → (q → p) (Another of the "Principles of transposition".)

✱2.18 (~p → p) → p (Called "The complement of reductio ad absurdum. It states that a proposition which follows from the hypothesis of its own falsehood is true" (PM, pp. 103–104).)

Most of these theorems—in particular ✱2.1, ✱2.11, and ✱2.14—are rejected by intuitionism. These tools are recast into another form that Kolmogorov cites as "Hilbert's four axioms of implication" and "Hilbert's two axioms of negation" (Kolmogorov in van Heijenoort, p. 335).

Propositions ✱2.12 and ✱2.14, "double negation": The intuitionist writings of L. E. J. Brouwer refer to what he calls "the principle of the reciprocity of the multiple species, that is, the principle that for every system the correctness of a property follows from the impossibility of the impossibility of this property" (Brouwer, ibid, p. 335).

This principle is commonly called "the principle of double negation" (PM, pp. 101–102). From the law of excluded middle (✱2.1 and ✱2.11), PM derives principle ✱2.12 immediately. We substitute ~p for p in 2.11 to yield ~p ∨ ~(~p), and by the definition of implication (i.e. 1.01 p → q = ~p ∨ q) then ~p ∨ ~(~p)= p → ~(~p). QED (The derivation of 2.14 is a bit more involved.)

Use in Computer Science Proofs

The law of excluded middle can be used to prove the decidability of certain computational problems. Usually, decidability is proved by showing an algorithm that solves the problem (i.e. a constructive proof). However, in some cases it is possible to prove that a problem is decidable without showing an algorithm that solves it.

For example, consider the following constant function f:

$$f = \begin{cases} 1 & \text{if Goldbach's conjecture is true} \\ 0 & \text{else} \end{cases}$$

By the law of excluded middle, Goldbach's conjecture is either true or false. If it is true then f is 1, and the required algorithm is just "print 1". If it is false then the required algorithm is just "print 0". In either case, there is a simple, one-line algorithm that prints f, so by definition, f is computationally decidable. It is true that we don"t know which algorithm to use, but we do know that an algorithm exists.

A slightly more complicated example is:

$$f(n) = \begin{cases} 1 & 0^n \text{ occurs in the decimal representation of } \pi \\ 0 & \text{else} \end{cases}$$

The function f is computable because, by the Law of Excluded Middle, there are only two possibilities to consider:

- For every positive integer n, the string 0^n appears in the decimal representation of {\displaystyle \pi }. In this case, the algorithm that always returns 1 is always correct.
- There is a largest integer N such that 0^N appears in the decimal representation of {\displaystyle \pi }. In this case the following algorithm (with the value hard-coded) is always correct:

Zeros-in-pi(n):

if (n > N) then return 0 else return 1

We have no idea which of these possibilities is correct, or what value of N is the right one in the second case. Nevertheless, one of these algorithms is guaranteed to be correct. Thus, there is an algorithm to decide whether a string of n zeros appears in π; ; the problem is decidable.

Criticisms

Many modern logic systems replace the law of excluded middle with the concept of negation as failure. Instead of a proposition either being true or false, a proposition is either true or not able to be proven true. These two dichotomies only differ in logical systems that are not complete. The principle of negation as failure is used as a foundation for autoepistemic logic, and is widely used in logic programming. In these systems, the programmer is free to assert the law of excluded middle as a true fact, but it is not built-in a priori into these systems.

Mathematicians such as L. E. J. Brouwer and Arend Heyting have also contested the usefulness of the law of excluded middle in the context of modern mathematics.

DOUBLE NEGATION

In propositional logic, **double negation** is the theorem that states that "If a statement is true, then it is not the case that the statement is not true." This is expressed by saying that a proposition A is logically equivalent to not (not-A), or by the formula A ≡ ~(~A) where the sign ≡ expresses logical equivalence and the sign ~ expresses negation.

Like the law of the excluded middle, this principle is considered to be a law of thought in classical logic, but it is disallowed by intuitionistic logic. The principle was stated as a theorem of propositional logic by Russell and Whitehead in Principia Mathematica as:

$*4 \cdot 13.\ \vdash .\ p \equiv \sim(\sim p)$

"This is the principle of double negation, i.e. a proposition is equivalent of the falsehood of its negation."

The principium contradictiones of modern logicians (particularly Leibniz and Kant) in the formula A is not not-A, differs entirely in meaning and application from the Aristotelian proposition [i.e. Law of Contradiction: not (A and not-A) i.e. ~(A & ~A), or not ((B is A) and (B is not-A))]. This latter refers to the relation between an affirmative and a negative judgment.

According to Aristotle, one judgment [B is judged to be an A] contradicts another [B is judged to be a not-A]. The later proposition [A is not not-A] refers to the relation between subject and predicate in a single judgment; the predicate contradicts the subject. Aristotle states that one judgment is false when another is true; the later writers [Leibniz and Kant] state that a judgment is in itself and absolutely false, because the predicate contradicts the subject. What the later writers desire is a principle from which it can be known whether certain propositions are in themselves true. From the Aristotelian proposition we cannot immediately infer the truth or falsehood of any particular proposition, but only the impossibility of believing both affirmation and negation at the same time.

Double Negative Elimination

Double negative elimination (also called double negation elimination, double negative introduction, double negation introduction, double negation, or negation elimination) are two valid rules of replacement. They are the inferences that if A is true, then not not-A is true and its converse, that, if not not-A is true, then A is true. The rule allows one to introduce or eliminate a negation from a logical proof. The rule is based on the equivalence of, for example, It is false that it is not raining. and It is raining.

The double negation introduction rule is:

$$P \Rightarrow \neg\neg P$$

and the *double negation elimination* rule is:

$$\neg\neg P \Rightarrow P$$

Where "⇒" is a metalogical symbol representing "can be replaced in a proof with."

Formal Notation

The double negation introduction rule may be written in sequent notation:

$$P \vdash \neg\neg P$$

The double negation elimination rule may be written as:

$\neg\neg P \vdash P$

In rule form:

$$\frac{P}{\neg\neg P}$$

and

$$\frac{\neg\neg P}{P}$$

or as a tautology (plain propositional calculus sentence):

$P \to \neg\neg P$

and

$\neg\neg P \to P$

These can be combined together into a single biconditional formula:

$\neg\neg P \leftrightarrow P.$

Since biconditionality is an equivalence relation, any instance of $\neg\neg$A in a well-formed formula can be replaced by A, leaving unchanged the truth-value of the well-formed formula.

Double negative elimination is a theorem of classical logic, but not of weaker logics such as intuitionistic logic and minimal logic. Because of their constructive character, a statement such as It's not the case that it's not raining is weaker than It"s raining. The latter requires a proof of rain, whereas the former merely requires a proof that rain would not be contradictory. (This distinction also arises in natural language in the form of litotes.) Double negation introduction is a theorem of both intuitionistic logic and minimal logic, as is $\neg\neg\neg A \vdash \neg A$.

In set theory also we have the negation operation of the complement which obeys this property: a set A and a set $(A^C)^C$ (where A^C represents the complement of A) are the same.

LAW OF NONCONTRADICTION

In classical logic, the **law of non-contradiction** (LNC) (or the **law of contradiction** (PM) or the **principle of non-contradiction** (PNC), or the **principle of contradiction**) is the second of the three classic laws of thought. It states that contradictory statements cannot both be true in the same sense at the same time, e.g. the two propositions "A is B" and "A is not B" are mutually exclusive.

The principle was stated as a theorem of propositional logic by Russell and Whitehead in Principia Mathematica as:

$\ast 3 \cdot 24. \quad \vdash . \sim (p . \sim p)$

The law of non-contradiction, along with its complement, the law of excluded middle (the third of the three classic laws of thought), are correlates of the law of identity (the first of the three laws). Because the law of identity partitions its logical Universe into exactly two parts, it creates a dichotomy wherein the two parts are "mutually exclusive" and "jointly exhaustive". The law of non-contradiction is merely an expression of the mutually exclusive aspect of that dichotomy, and the law of excluded middle, an expression of its jointly exhaustive aspect.

Interpretations

One difficulty in applying the law of non-contradiction is ambiguity in the propositions. For instance, if time is not explicitly specified as part of the propositions A and B, then A may be B at one time, and not at another. A and B may in some cases be made to sound mutually exclusive linguistically even though A may be partly B and partly not B at the same time. However, it is impossible to predicate of the same thing, at the same time, and in the same sense, the absence and the presence of the same fixed quality.

Heraclitus

According to both Plato and Aristotle, Heraclitus was said to have denied the law of non-contradiction. This is quite likely if, as Plato pointed out, the law of non-contradiction does not hold for changing things in the world. If a philosophy of Becoming is not possible without change, then (the potential of) what is to become must already exist in the present object. In "We step and do not step into the same rivers; we are and we are not", both Heraclitus's and Plato's object simultaneously must, in some sense, be both what it now is and have the potential (dynamis) of what it might become.

Unfortunately, so little remains of Heraclitus' aphorisms that not much about his philosophy can be said with certainty. He seems to have held that strife of opposites is universal both within and without, therefore both opposite existents or qualities must simultaneously exist, although in some instances in different respects. "The road up and down are one and the same" implies either the road leads both ways, or there can be no road at all. This is the logical complement of the law of non-contradiction. According to Heraclitus, change, and the constant conflict of opposites is the universal logos of nature.

Protagoras

Personal subjective perceptions or judgments can only be said to be true at the same time in the same respect, in which case, the law of non-contradiction must be applicable to personal judgments. The most famous saying of Protagoras is: "Man is the measure of all things: of things which are, that they are, and of things which are not, that they are not". However, Protagoras was referring to things that are used by or in some way related to humans. This makes a great difference in the meaning of his aphorism. Properties, social entities, ideas, feelings, judgements, etc. originate in the human mind. However, Protagoras has never suggested that man must be the measure of stars, or the motion of the stars.

Parmenides

Parmenides employed an ontological version of the law of non-contradiction to prove that being is and to deny the void, change, and motion. He also similarly disproved contrary propositions. In his poem On Nature, he said, the only routes of inquiry there are for thinking: the one that [it] is and that [it] cannot not be is the path of Persuasion (for it attends upon truth) the other, that [it] is not and that it is right that [it] not be, this I point out to you is a path wholly inscrutable for you could not know what is not (for it is not to be accomplished) nor could you point it out... For the same thing is for thinking and for being The nature of the 'is' or what-is in Parmenides is a highly contentious subject. Some have taken it to be whatever exists, some to be whatever is or can be the object of scientific inquiry.

Socrates

In Plato's early dialogues, Socrates uses the elenctic method to investigate the nature or definition of ethical concepts such as justice or virtue. Elenctic refutation depends on a dichotomous thesis, one that may be divided into exactly two mutually exclusive parts, only one of which may be true. Then Socrates goes on to demonstrate the contrary of the commonly accepted part using the law of non-contradiction. According to Gregory Vlastos, the method has the following steps:

- Socrates" interlocutor asserts a thesis, for example "Courage is endurance of the soul", which Socrates considers false and targets for refutation.
- Socrates secures his interlocutor"s agreement to further premises, for example "Courage is a fine thing" and "Ignorant endurance is not a fine thing".
- Socrates then argues, and the interlocutor agrees, that these further premises imply the contrary of the original thesis, in this case it leads to: "courage is not endurance of the soul".
- Socrates then claims that he has shown that his interlocutor"s thesis is false and that its negation is true.

Plato's synthesis

Plato's version of the law of non-contradiction states that "The same thing clearly cannot act or be acted upon in the same part or in relation to the same thing at the same time, in contrary ways" (The Republic (436b)). In this, Plato carefully phrases three axiomatic restrictions on action or reaction: 1) in the same part, 2) in the same relation, 3) at the same time. The effect is to momentarily create a frozen, timeless state, somewhat like figures frozen in action on the frieze of the Parthenon.

This way, he accomplishes two essential goals for his philosophy. First, he logically separates the Platonic world of constant change from the formally knowable world of momentarily fixed physical objects. Second, he provides the conditions for the dialectic method to be used in finding definitions, as for example in the Sophist. So Plato's law of non-contradiction is the empirically derived necessary starting point for all else he has to say. In contrast, Aristotle reverses Plato's order of derivation. Rather than starting with experience, Aristotle begins a priori with the law of non-contradiction as the fundamental axiom of an analytic philosophical system. This axiom then necessitates the fixed, realist model. Now, he starts with much stronger logical foundations than Plato's non-contrariety of action in reaction to conflicting demands from the three parts of the soul.

Aristotle"s contribution

The traditional source of the law of non-contradiction is Aristotle's Metaphysics where he gives three different versions.

- ontological: "It is impossible that the same thing belong and not belong to the same thing at the same time and in the same respect." (1005b19-20)
- psychological: "No one can believe that the same thing can (at the same time) be and not be." (1005b23-24)
- logical: "The most certain of all basic principles is that contradictory propositions are not true simultaneously." (1011b13-14)

Aristotle attempts several proofs of this law. He first argues that every expression has a single meaning (otherwise we could not communicate with one another). This rules out the possibility that by "to be a man", "not to be a man" is meant. But "man" means "two-footed animal" (for example), and so if anything is a man, it is necessary (by virtue of the meaning of "man") that it must be a two-footed animal, and so it is impossible at the same time for it not to be a two-footed animal. Thus "it is not possible to say truly at the same time that the same thing is and is not a man" (Metaphysics 1006b 35). Another argument is that anyone who believes something cannot believe its contradiction (1008b).

Why does he not just get up first thing and walk into a well or, if he finds one, over a cliff? In fact, he seems rather careful about cliffs and wells.

Avicenna

Avicenna, a Persian philosopher, said, "Anyone who denies the law of non-contradiction should be beaten and burned until he admits that to be beaten is not the same as not to be beaten, and to be burned is not the same as not to be burned."

Eastern philosophy

The law of non-contradiction is found in ancient Indian logic as a meta-rule in the Shrauta Sutras,the grammar of Pānini, and the Brahma Sutras attributed to Vyasa. It was later elaborated on by medieval commentators such as Madhvacharya.

Leibniz and Kant

Leibniz and Kant adopted a different statement, by which the law assumes an essentially different meaning. Their formula is A is not not-A; in other words it is impossible to predicate of a thing a quality which is its contradictory. Unlike Aristotle's law this law deals with the necessary relation between subject and predicate in a single judgment. For example, in Gottlob Ernst Schulze"s Aenesidemus, it is asserted, "... nothing supposed capable of being thought may contain contradictory characteristics." Whereas Aristotle states that one or other of two contradictory propositions must be false, the Kantian law states that a particular kind of proposition is in itself necessarily false. On the other hand, there is a real connection between the two laws. The denial of the statement A is not-A presupposes some knowledge of what A is, i.e. the statement A is A. In other words, a judgment about A is implied.

Kant's analytical judgments of propositions depend on presupposed concepts which are the same for all people. His statement, regarded as a logical principle purely and apart from material facts, does not therefore amount to more than that of Aristotle, which deals simply with the significance of negation.

Modern Logics

Traditionally, in Aristotle's classical logical calculus, in evaluating any proposition there are only two possible truth values, "true" and "false." An obvious extension to classical two-valued logic is a many-valued logic for more than two possible values. In logic, a many- or multi-valued logic is a propositional calculus in which there are more than two values. Those most popular in the literature are three-valued (e.g., Łukasiewicz's and Kleene"s), which accept the values "true", "false", and "unknown", finite-valued with more than three values, and the infinite-valued (e.g. fuzzy logic and probability logic) logics.

Dialetheism

Graham Priest advocates the view that under some conditions, some statements can be both true and false simultaneously, or may be true and false at different times. Dialetheism arises from formal logical paradoxes, such as the Liar's paradox and Russell"s paradox.

Alleged Impossibility of its Proof or Denial

As is true of all axioms of logic, the law of non-contradiction is alleged to be neither verifiable nor falsifiable, on the grounds that any proof or disproof must use the law itself prior to reaching the conclusion. In other words, in order to verify or falsify the laws of logic one must resort to logic as a weapon, an act which would essentially be self-defeating. Since the early 20th century, certain logicians have proposed logics that deny the validity of the law. Collectively, these logics are known as "paraconsistent" or "inconsistency-tolerant" logics. But not all paraconsistent logics deny the law, since they are not necessarily completely agnostic to inconsistencies in general. Graham Priest advances the strongest thesis of this sort, which he calls "dialetheism".

In several axiomatic derivations of logic, this is effectively resolved by showing that $(P \vee \neg P)$ and its negation are constants, and simply defining TRUE as $(P \vee \neg P)$ and FALSE as $\neg(P \vee \neg P)$, without taking a position as to the principle of bivalence or the law of excluded middle.

Some, such as David Lewis, have objected to paraconsistent logic on the ground that it is simply impossible for a statement and its negation to be jointly true. A related objection that is that "negation" in paraconsistent logic is not really negation; it is merely a subcontrary-forming operator.

Principle of Explosion

The **principle of explosion** (Latin: ex falso (sequitur) quodlibet (EFQ), "from falsehood, anything (follows)", or ex contradictione (sequitur) quodlibet (ECQ), "from contradiction, anything (follows)"), or the **principle of Pseudo-Scotus**, is the law of classical logic, intuitionistic logic and similar logical systems, according to which any statement can be proven from a contradiction. That is, once a contradiction has been asserted, any proposition (including their negations) can be inferred from it.

As a demonstration of the principle, consider two contradictory statements – "All lemons are yellow" and "Not all lemons are yellow", and suppose (for the sake of argument) that both are simultaneously true. If that is the case, anything can be proven, e.g. "unicorns exist", by using the following argument:

- We know that "All lemons are yellow" as it is defined to be true.
- Therefore, the statement that ("All lemons are yellow" OR "unicorns exist") must also be true, since the first part is true.
- However, if "Not all lemons are yellow" (and this is also defined to be true), unicorns must exist – otherwise statement 2 would be false. It has thus been "proven" that unicorns exist. The same could be applied to any assertion, including the statement "unicorns do not exist".

The principle is not a universal rule; rather it exists as a consequence of a choice of which logic to use. It does not appear in some paraconsistent logics which allow localised ‹gluts" of contradictory statements to be proved without affecting other proofs. In artificial intelligence and models of human reasoning it is common for such logics to be used. Truth maintenance systems are AI models which try to capture this process.

Symbolic Representation

The principle of explosion can be expressed in the following way

$$\forall P \forall Q : (P \wedge \neg P) \vdash Q$$

(where symbolizes the relation of logical consequence and symbolizes a contradiction) :

$$\{\phi, \neg\phi\} \vdash \psi$$

or

$$\bot \to P.$$

This can be read as, "If one claims something (ϕ) and its negation $(\neg\phi)$, one can logically derive any conclusion (ψ)."

Arguments for Explosion

An informal, descriptive, argument is given above. In more formal terms, there are two kinds of argument for the principle of explosion, semantic and proof-theoretic.

The Semantic Argument

The first argument is semantic or model-theoretic in nature. A sentence ψ is a semantic consequence of a set of sentences Γ only if every model of Γ is a model of ψ. But there is no model of the contradictory set $\{\phi, \neg\phi\}$. A fortiori, there is no model of $\{\phi, \neg\phi\}$ that is not a model of ψ. Thus, vacuously, every model of $\{\phi, \neg\phi\}$ is a model of ψ. Thus ψ is a semantic consequence of $\{\phi, \neg\phi\}$.

Addressing the Principle

Paraconsistent logics have been developed that allow for sub-contrary forming operators. Model-theoretic paraconsistent logicians often deny the assumption that there can be no model of $\{\phi, \neg\phi\}$ and devise semantical systems in which there are such models. Alternatively, they reject the idea that propositions can be classified as true or false. Proof-theoretic paraconsistent logics usually deny the validity of one of the steps necessary for deriving an explosion, typically including disjunctive syllogism, disjunction introduction, and reductio ad absurdum.

Use

The metamathematical value of the principle of explosion is that for any logical system where this principle holds, any derived theory which proves ⊥ (or an equivalent form, $\phi \wedge \neg\phi$) is worthless because all its statements would become theorems, making it impossible to distinguish truth from falsehood. That is to say, the principle of explosion is an argument for the law of non-contradiction in classical logic, because without it all truth statements become meaningless.

MONOTONICITY OF ENTAILMENT

Monotonicity of entailment is a property of many logical systems that states that the hypotheses of any derived fact may be freely extended with additional assumptions. In sequent calculi this property can be captured by an inference rule called **weakening**, or sometimes **thinning**, and in such systems one may say that entailment is monotone if and only if the rule is admissible. Logical systems with this property are occasionally called monotonic logics in order to differentiate them from non-monotonic logics.

Weakening rule

To illustrate, starting from the natural deduction sequent:

$\Gamma \vdash C$

weakening allows one to conclude:

$\Gamma, A \vdash C$

For example, the syllogism "All men are mortal. Socrates is a man. Therefore Socrates is mortal." can be weakened by adding a premise: "All men are mortal. Socrates is a man. Cows produce milk. Therefore Socrates is mortal." The validity of the original conclusion is not changed by the addition of premises.

Non-monotonic Logics

In most logics, weakening is either an inference rule or a metatheorem if the logic doesn't have an explicit rule. Notable exceptions are:

- Strict logic or relevant logic, where every hypothesis must be necessary for the conclusion.
- Linear logic which disallows arbitrary contraction in addition to arbitrary weakening.
- Bunched implications where weakening is restricted to additive composition.
- Various types of default reasoning.
- Abductive reasoning, the process of deriving the most likely explanations of the known facts.
- Reasoning about knowledge, where statements specifying that something is not known need to be retracted when that thing is learned.

COMMUTATIVITY OF CONJUNCTION

In propositional logic, the **commutativity of conjunction** is a valid argument form and truth-functional tautology. It is considered to be a law of classical logic. It is the principle that the conjuncts of a logical conjunction may switch places with each other, while preserving the truth-value of the resulting proposition.

Formal Notation

Commutativity of conjunction can be expressed in sequent notation as:

$$(P \land Q) \vdash (Q \land P)$$

and

$$(Q \land P) \vdash (P \land Q)$$

where $\vdash$ is a metalogical symbol meaning that $(Q \land P)$ is a syntactic consequence of , in the one case, and $(P \land Q)$ is a syntactic consequence of $(Q \land P)$ in the other, in some logical system; or in rule form:

$$\frac{P \land Q}{\therefore Q \land P}$$

and

$$\frac{Q \land P}{\therefore P \land Q}$$

where the rule is that wherever an instance of $(P \land Q)$" appears on a line of a proof, it can be replaced with "$(Q \land P)$" and wherever an instance of "$(Q \land P)$" appears on a line of a proof, it can be replaced with " $(P \land Q)$";

or as the statement of a truth-functional tautology or theorem of propositional logic:

$$(P \wedge Q) \rightarrow (Q \wedge P)$$

and

$$(Q \wedge P) \rightarrow (P \wedge Q)$$

where P and Q are propositions expressed in some formal system.

Generalized Principle

For any propositions H_1, H_2, ... H_n, and permutation σ(n) of the numbers 1 through n, it is the case that:

$$H_1 \wedge H_2 \wedge \ldots \wedge H_n$$

is equivalent to

$$H_{\sigma(1)} \wedge H_{\sigma(2)} \wedge H_{\sigma(n)}.$$

For example, if H_1 is

It is raining

H_2 is

Socrates is mortal

and H_3 is

2+2=4

then

It is raining and Socrates is mortal and 2+2=4

is equivalent to

Socrates is mortal and 2+2=4 and it is raining

and the other orderings of the predicates.

DE MORGAN'S LAWS

In propositional logic and boolean algebra, **De Morgan's laws** are a pair of transformation rules that are both validrules of inference. They are named after Augustus De Morgan, a 19th-century British mathematician. The rules allow the expression of conjunctions and disjunctions purely in terms of each other via negation.

The rules can be expressed in English as:

the negation of a conjunction is the disjunction of the negations; and the negation of a disjunction is the conjunction of the negations; or the complement of the union of two sets is the same as the intersection of their complements; and the complement of the intersection of two sets is the same as the union of their complements.

In set theory and Boolean algebra, these are written formally as

$$\overline{A \cup B} = \overline{A} \cap \overline{B},$$
$$\overline{A \cap B} = \overline{A} \cup \overline{B},$$

where

- A and B are sets,
- A is the complement of A,
- ∩ is the intersection, and
- ∪ is the union.

In formal language, the rules are written as

$$\neg(P \wedge Q) \iff (\neg P) \vee (\neg Q),$$

and

$$\neg(P \vee Q) \iff (\neg P) \wedge (\neg Q),$$

where

- *P* and *Q are propositions,*
- ¬ is the negation logic operator (NOT),
- ∧ is the conjunction logic operator (AND),
- ∨ is the disjunction logic operator (OR),
- $\iff$ is a metalogical symbol meaning "can be replaced in a logical proof with".

Applications of the rules include simplification of logical expressions in computer programs and digital circuit designs. De Morgan's laws are an example of a more general concept of mathematical duality.

Formal notation

The negation of conjunction rule may be written in sequent notation:

$$\neg(P \wedge Q) \vdash (\neg P \vee \neg Q).$$

The negation of disjunction rule may be written as:

$$\neg(P \vee Q) \vdash (\neg P \wedge \neg Q).$$

In rule form: *negation of conjunction*

$$\frac{\neg(P \wedge Q)}{\therefore \neg P \vee \neg Q}$$

and *negation of disjunction*

$$\frac{\neg(P \vee Q)}{\therefore \neg P \wedge \neg Q}$$

and expressed as a truth-functional tautology or theorem of propositional logic:

$$\neg(P \wedge Q) \to (\neg P \vee \neg Q),$$
$$\neg(P \vee Q) \to (\neg P \wedge \neg Q),$$

where P and Q are propositions expressed in some formal system.

Substitution form

De Morgan's laws are normally shown in the compact form above, with negation of the output on the left and negation of the inputs on the right. A clearer form for substitution can be stated as:

$$(P \wedge Q) \equiv \neg(\neg P \vee \neg Q),$$
$$(P \vee Q) \equiv \neg(\neg P \wedge \neg Q).$$

This emphasizes the need to invert both the inputs and the output, as well as change the operator, when doing a substitution.

Set theory and Boolean algebra

In set theory and Boolean algebra, it is often stated as "union and intersection interchange under complementation", which can be formally expressed as:

$$\overline{A \cup B} = \overline{A} \cap \overline{B},$$
$$\overline{A \cap B} = \overline{A} \cup \overline{B},$$

where:

- A is the negation of A, the overline being written above the terms to be negated,
- ∩ is the intersection operator (AND),
- ∪ is the union operator (OR).

The generalized form is:

$$\overline{\bigcap_{i \in I} A_i} \equiv \bigcup_{i \in I} \overline{A_i},$$
$$\overline{\bigcup_{i \in I} A_i} = \bigcap_{i \in I} \overline{A_i},$$

where I is some, possibly uncountable, indexing set.

In set notation, De Morgan"s laws can be remembered using the mnemonic "break the line, change the sign".

Engineering

In electrical and computer engineering, De Morgan's laws are commonly written as:

$$\overline{A \cdot B} \equiv \overline{A} + \overline{B}$$

and

$$\overline{A + B} \equiv \overline{A} \cdot \overline{B},$$

where:

- . is a logical AND,
- +is a logical OR,
- the overbar is the logical NOT of what is underneath the overbar.

Text Searching

De Morgan's laws commonly apply to text searching using Boolean operators AND, OR, and NOT. Consider a set of documents containing the words "cars" and "trucks". De Morgan's laws hold that these two searches will return the same set of documents:

Search A: NOT (cars OR trucks)

Search B: (NOT cars) AND (NOT trucks)

The corpus of documents containing "cars" or "trucks" can be represented by four documents:

Document 1: Contains only the word "cars".

Document 2: Contains only "trucks".

Document 3: Contains both "cars" and "trucks".

Document 4: Contains neither "cars" nor "trucks".

To evaluate Search A, clearly the search "(cars OR trucks)" will hit on Documents 1, 2, and 3. So the negation of that search (which is Search A) will hit everything else, which is Document 4.

Evaluating Search B, the search "(NOT cars)" will hit on documents that do not contain "cars", which is Documents 2 and 4. Similarly the search "(NOT trucks)" will hit on Documents 1 and 4. Applying the AND operator to these two searches (which is Search B) will hit on the documents that are common to these two searches, which is Document 4.

A similar evaluation can be applied to show that the following two searches will return the same set of documents (Documents 1, 2, 4):

Search C: NOT (cars AND trucks),

Search D: (NOT cars) OR (NOT trucks).

The laws are named after Augustus De Morgan (1806–1871), who introduced a formal version of the laws to classical propositional logic. De Morgan"s formulation was influenced by algebraization of logic undertaken by George Boole, which later cemented De Morgan"s claim to the find. Nevertheless, a similar observation was made by Aristotle, and was known to Greek and Medieval logicians. For example, in the 14th century, William of Ockham wrote down the words that would result by reading the laws out. Jean Buridan, in his Summulae de Dialectica, also describes rules of conversion that follow the lines of De Morgan's laws. Still, De Morgan is given credit for stating the laws in the terms of modern formal logic, and incorporating them into the language of logic. De Morgan's laws can be proved easily, and may even seem trivial. Nonetheless, these laws are helpful in making valid inferences in proofs and deductive arguments.

Informal proof

De Morgan's theorem may be applied to the negation of a disjunction or the negation of a conjunction in all or part of a formula.

Negation of a Disjunction

In the case of its application to a disjunction, consider the following claim: "it is false that either of A or B is true", which is written as:

$\neg(A \vee B)$.

In that it has been established that neither A nor B is true, then it must follow that both A is not true and B is not true, which may be written directly as:

$(\neg A) \wedge (\neg B)$

If either A or B were true, then the disjunction of A and B would be true, making its negation false. Presented in English, this follows the logic that "since two things are both false, it is also false that either of them is true".

Working in the opposite direction, the second expression asserts that A is false and B is false (or equivalently that "not A" and "not B" are true). Knowing this, a disjunction of A and B must be false also. The negation of said disjunction must thus be true, and the result is identical to the first claim.

Negation of a Conjunction

The application of De Morgan's theorem to a conjunction is very similar to its application to a disjunction both in form and rationale. Consider the following claim: "it is false that A and B are both true", which is written as:

$\neg(A \wedge B)$

In order for this claim to be true, either or both of A or B must be false, for if they both were true, then the conjunction of A and B would be true, making its negation false. Thus, one (at least) or more of A and B must be false (or equivalently, one or more of "not A" and "not B" must be true). This may be written directly as,

$(\neg A) \vee (\neg B)$

Presented in English, this follows the logic that "since it is false that two things are both true, at least one of them must be false".

Working in the opposite direction again, the second expression asserts that at least one of "not A" and "not B" must be true, or equivalently that at least one of A and B must be false. Since at least one of them must be false, then their conjunction would likewise be false. Negating said conjunction thus results in a true expression, and this expression is identical to the first claim.

Formal Proof

The proof that $(A \cap B)^c = A^c \cup B^c$ is completed in 2 steps by proving both $(A \cap B)^c \subseteq A^c \cup B^c$ and $A^c \cup B^c \subseteq (A \cap B)^c$.

Part 1

Let $x \in (A \cap B)^c$. Then, $x \notin A \cap B$.

Because $A \cap B = \{y | y \in A \land y \in B\}$, it must be the case that $x \notin A$ or $x \notin B$.

If $x \notin A$, then $x \in A^c$, so $x \in A^c \cup B^c$.

Similarly, if $x \notin B$, then $x \in B^c$, so $x \in A^c \cup B^c$.

Thus, $\forall x (x \in (A \cap B)^c \to x \in A^c \cup B^c)$;

that is, $(A \cap B)^c \subseteq A^c \cup B^c$.

Part 2

To prove the reverse direction, let $x \in A^c \cup B^c$, and assume $x \notin (A \cap B)^c$.

Under that assumption, it must be the case that $x \in A \cap B$,

so it follows that $x \in A$ and $x \in B$, and thus $x \notin A^c$ and $x \notin B^c$.

However, that means $x \notin A^c \cup B^c$, in contradiction to the hypothesis that $x \in A^c \cup B^c$,

therefore, the assumption $x \notin (A \cap B)^c$ must not be the case, meaning that $x \in (A \cap B)^c$.

Hence, $\forall x (x \in A^c \cup B^c \to x \in (A \cap B)^c)$,

that is, $A^c \cup B^c \subseteq (A \cap B)^c$.

Conclusion

If $A^c \cup B^c \subseteq (A \cap B)^c$ and $(A \cap B)^c \subseteq A^c \cup B^c$, then $(A \cap B)^c = A^c \cup B^c$; this concludes the proof of De Morgan's law.

The other De Morgan's law, $(A \cup B)^c = A^c \cap B^c$, is proven similarly.

Generalising De Morgan duality

In extensions of classical propositional logic, the duality still holds (that is, to any logical operator one can always find its dual), since in the presence of the identities governing negation, one may always introduce an operator that is the De Morgan dual of another. This leads to an important property of logics based on classical logic, namely the existence of negation normal forms: any formula is equivalent to another formula where negations only occur applied to the non-logical atoms of the formula. The existence of negation normal forms drives many applications, for example in digital circuit design, where it is used to manipulate the types of logic gates, and in formal logic, where it is needed to find the conjunctive normal form and disjunctive normal form of a formula. Computer programmers use them to simplify or properly negate complicated logical conditions. They are also often useful in computations in elementary probability theory.

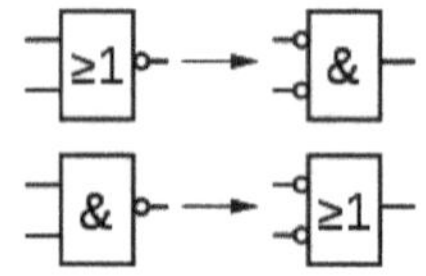

De Morgan's Laws represented as a circuit with logic gates

Let one define the dual of any propositional operator P(p, q, ...) depending on elementary propositions p, q, ... to be the operator P^d defined by

$$P^d(p, q, \ldots) = \neg P(\neg p, \neg q, \ldots)$$

Extension to Predicate and Modal Logic

This duality can be generalised to quantifiers, so for example the universal quantifier and existential quantifier are duals:

$$\forall x\, P(x) \equiv \neg[\exists x\, \neg P(x)]$$
$$\exists x\, P(x) \equiv \neg[\forall x\, \neg P(x)]$$

To relate these quantifier dualities to the De Morgan laws, set up a model with some small number of elements in its domain D, such as

D = {a, b, c}.

Then

$$\forall x\, P(x) \equiv P(a) \wedge P(b) \wedge P(c)$$

and

$$\exists x\, P(x) \equiv P(a) \vee P(b) \vee P(c).$$

But, using De Morgan's laws,

$$P(a) \wedge P(b) \wedge P(c) \equiv \neg(\neg P(a) \vee \neg P(b) \vee \neg P(c))$$

and

$$P(a) \vee P(b) \vee P(c) \equiv \neg(\neg P(a) \wedge \neg P(b) \wedge \neg P(c)),$$

verifying the quantifier dualities in the model.

Then, the quantifier dualities can be extended further to modal logic, relating the box ("necessarily") and diamond ("possibly") operators:

$$\Box p \equiv \neg \Diamond \neg p,$$
$$\Diamond p \equiv \neg \Box \neg p.$$

In its application to the alethic modalities of possibility and necessity, Aristotle observed this case, and in the case of normal modal logic, the relationship of these modal operators to the quantification can be understood by setting up models using Kripke semantics.

Propositional Calculus

Propositional calculus (also called propositional logic, sentential calculus, sentential logic, or sometimes zeroth-order logic) is the branch of logic concerned with the study of propositions (whether they are true or false) that are formed by other propositions with the use of logical connectives, and how their value depends on the truth value of their components. Logical connectives are found in natural languages. In English for example, some examples are "and" (conjunction), "or" (disjunction), "not" (negation) and "if" (but only when used to denote material conditional).

The following is an example of a very simple inference within the scope of propositional logic:

Premise 1: If it"s raining then it"s cloudy.

Premise 2: It"s raining.

Conclusion: It"s cloudy.

Both premises and the conclusion are propositions. The premises are taken for granted and then with the application of modus ponens (an inference rule) the conclusion follows.

As propositional logic is not concerned with the structure of propositions beyond the point where they can"t be decomposed anymore by logical connectives, this inference can be restated replacing those atomic statements with statement letters, which are interpreted as variables representing statements:

Premise 1: $P \to Q$

Premise 2: P

Conclusion: Q

The same can be stated succinctly in the following way:

$$P \to Q, P \vdash Q$$

When P is interpreted as "It's raining" and Q as "it's cloudy" the above symbolic expressions can be seen to exactly correspond with the original expression in natural language. Not only that, but they will also correspond with any other inference of this form, which will be valid on the same basis that this inference is.

Propositional logic may be studied through a formal system in which formulas of a formal language may be interpreted to represent propositions. A system of inference rules and axioms allows certain formulas to be derived. These derived formulas are called theorems and may be interpreted to be true propositions. A constructed sequence of such formulas is known as a derivation or proof and the last formula of the sequence is the theorem. The derivation may be interpreted as proof of the proposition represented by the theorem.

When a formal system is used to represent formal logic, only statement letters are represented directly. The natural language propositions that arise when they're interpreted are outside the scope of the system, and the relation between the formal system and its interpretation is likewise outside the formal system itself.

Usually in **truth-functional propositional logic**, formulas are interpreted as having either a truth value of true or a truth value of false. Truth-functional propositional logic and systems isomorphic to it, are considered to be **zeroth-order logic**.

Although propositional logic (which is interchangeable with propositional calculus) had been hinted by earlier philosophers, it was developed into a formal logic by Chrysippus in the 3rd century BC and expanded by the Stoics. The logic was focused on propositions. This advancement was different from the traditional syllogistic logic which was focused on terms. However, later in antiquity, the propositional logic developed by the Stoics was no longer understood[who?]. Consequently, the system was essentially reinvented by Peter Abelard in the 12th century.

Propositional logic was eventually refined using symbolic logic. The 17th/18th-century mathematician Gottfried Leibniz has been credited with being the founder of symbolic logic for his work with the calculus ratiocinator. Although his work was the first of its kind, it was unknown to the larger logical community. Consequently, many of the advances achieved by Leibniz were reachieved by logicians like George Boole and Augustus De Morgan completely independent of Leibniz.

Just as propositional logic can be considered an advancement from the earlier syllogistic logic, Gottlob Frege"s predicate logic was an advancement from the earlier propositional logic. One author describes predicate logic as combining "the distinctive features of syllogistic logic and propositional logic."Consequently, predicate logic ushered in a new era in logic"s history; however, advances in propositional logic were still made after Frege, including Natural Deduction, Truth-Trees and Truth-Tables. Natural deduction was invented by Gerhard Gentzen and Jan Łukasiewicz. Truth-Trees were invented by Evert Willem Beth. The invention of truth-tables, however, is of controversial attribution.

Within works by Frege and Bertrand Russell, one finds ideas influential in bringing about the notion of truth tables. The actual ‹tabular structure" (being formatted as a table), itself, is generally credited to either Ludwig Wittgenstein or Emil Post (or both, independently). Besides Frege and Russell, others credited with having ideas preceding truth-tables include Philo, Boole, Charles Sanders Peirce, and Ernst Schröder. Others credited with the tabular structure include Łukasiewicz, Schröder, Alfred North Whitehead, William Stanley Jevons, John Venn, and Clarence Irving Lewis. Ultimately, some have concluded, like John Shosky, that "It is far from clear that any one person should be given the title of ‹inventor" of truth-tables.".

Terminology

In general terms, a calculus is a formal system that consists of a set of syntactic expressions (well-formed formulas), a distinguished subset of these expressions (axioms), plus a set of formal rules that define a specific binary relation, intended to be interpreted as logical equivalence, on the space of expressions.

When the formal system is intended to be a logical system, the expressions are meant to be interpreted as statements, and the rules, known to be inference rules, are typically intended to be truth-preserving. In this setting, the rules (which may include axioms) can then be used to derive ("infer") formulas representing true statements from given formulas representing true statements.

The set of axioms may be empty, a nonempty finite set, a countably infinite set, or be given by axiom schemata. A formal grammar recursively defines the expressions and well-formed formulas of the language. In addition a semantics may be given which defines truth and valuations (or interpretations).

The language of a propositional calculus consists of

- a set of primitive symbols, variously referred to as atomic formulas, placeholders, proposition letters, or variables, and
- a set of operator symbols, variously interpreted as logical operators or logical connectives.

A well-formed formula is any atomic formula, or any formula that can be built up from atomic formulas by means of operator symbols according to the rules of the grammar.

Mathematicians sometimes distinguish between propositional constants, propositional variables, and schemata. Propositional constants represent some particular proposition, while propositional variables range over the set of all atomic propositions. Schemata, however, range over all propositions. It is common to represent propositional constants by A, B, and C, propositional variables by P, Q, and R, and schematic letters are often Greek letters, most often φ, ψ, and χ.

Basic concepts

The following outlines a standard propositional calculus. Many different formulations exist which are all more or less equivalent but differ in the details of: their language, that is, the particular collection of primitive symbols and operator symbols,

- the set of axioms, or distinguished formulas, and
- the set of inference rules.

Any given proposition may be represented with a letter called a ‹propositional constant", analogous to representing a number by a letter in mathematics, for instance, a = 5. All propositions require exactly one of two truth-values: true or false. For example, let P be the proposition that it is raining outside. This will be true (P) if it is raining outside and false otherwise (¬P).

- We then define truth-functional operators, beginning with negation. ¬P represents the negation of P, which can be thought of as the denial of P. In the example above, ¬P expresses that it is not raining outside, or by a more standard reading: "It is not the case that it is raining outside." When P is true, ¬P is false; and when P is false, ¬P is true. ¬¬P always has the same truth-value as P.
- Conjunction is a truth-functional connective which forms a proposition out of two simpler propositions, for example, P and Q. The conjunction of P and Q is written P ∧ Q, and expresses that each are true. We read P ∧ Q for "P and Q". For any two propositions, there are four possible assignments of truth values:
    - P is true and Q is true
    - P is true and Q is false
    - P is false and Q is true
    - P is false and Q is false

The conjunction of P and Q is true in case 1 and is false otherwise. Where P is the proposition that it is raining outside and Q is the proposition that a cold-front is over Kansas, P ∧ Q is true when it is raining outside and there is a cold-front over Kansas. If it is not raining outside, then P ∧ Q is false; and if there is no cold-front over Kansas, then P ∧ Q is false.

- Disjunction resembles conjunction in that it forms a proposition out of two simpler propositions. We write it P ∨ Q, and it is read "P or Q". It expresses that either P or Q is true. Thus, in the cases listed above, the disjunction of P and Q is true in all cases except 4. Using the example above, the disjunction expresses that it is either raining outside or there is a cold front over Kansas. (Note, this use of disjunction is supposed to resemble the use of the English word "or". However, it is most like the English inclusive "or", which can be used to express the truth of at least one of two propositions. It is not like the English exclusive "or", which expresses the truth of exactly one of two propositions. That is to say, the exclusive "or" is false when both P and Q are true (case 1). An example of the exclusive or is: You may have a bagel or a pastry, but not both. Often in natural language, given the appropriate context, the addendum "but not both" is omitted but implied. In mathematics, however, "or" is always inclusive or; if exclusive or is meant it will be specified, possibly by "xor".)

- Material conditional also joins two simpler propositions, and we write $P \rightarrow Q$, which is read "if P then Q". The proposition to the left of the arrow is called the antecedent and the proposition to the right is called the consequent. (There is no such designation for conjunction or disjunction, since they are commutative operations.) It expresses that Q is true whenever P is true. Thus it is true in every case above except case 2, because this is the only case when P is true but Q is not. Using the example, if P then Q expresses that if it is raining outside then there is a cold-front over Kansas. The material conditional is often confused with physical causation. The material conditional, however, only relates two propositions by their truth-values—which is not the relation of cause and effect. It is contentious in the literature whether the material implication represents logical causation.
- Biconditional joins two simpler propositions, and we write $P \leftrightarrow Q$, which is read "P if and only if Q". It expresses that P and Q have the same truth-value, thus P if and only if Q is true in cases 1 and 4, and false otherwise.

It is extremely helpful to look at the truth tables for these different operators, as well as the method of analytic tableaux.

Closure under Operations

Propositional logic is closed under truth-functional connectives. That is to say, for any proposition φ, $\neg\varphi$ is also a proposition. Likewise, for any propositions φ and ψ, $\varphi \land \psi$ is a proposition, and similarly for disjunction, conditional, and biconditional. This implies that, for instance, $\varphi \land \psi$ is a proposition, and so it can be conjoined with another proposition. In order to represent this, we need to use parentheses to indicate which proposition is conjoined with which. For instance, $P \land Q \land R$ is not a well-formed formula, because we do not know if we are conjoining $P \land Q$ with R or if we are conjoining P with $Q \land R$. Thus we must write either $(P \land Q) \land R$ to represent the former, or $P \land (Q \land R)$ to represent the latter. By evaluating the truth conditions, we see that both expressions have the same truth conditions (will be true in the same cases), and moreover that any proposition formed by arbitrary conjunctions will have the same truth conditions, regardless of the location of the parentheses. This means that conjunction is associative, however, one should not assume that parentheses never serve a purpose. For instance, the sentence $P \land (Q \lor R)$ does not have the same truth conditions of $(P \land Q) \lor R$, so they are different sentences distinguished only by the parentheses. One can verify this by the truth-table method referenced above.

Note: For any arbitrary number of propositional constants, we can form a finite number of cases which list their possible truth-values. A simple way to generate this is by truth-tables, in which one writes P, Q, ..., Z, for any list of k propositional constants—that is to say, any list of propositional constants with kentries. Below this list, one writes 2^k rows, and below P one fills in the first half of the rows with true (or T) and the second half with false (or F). Below Q one fills in one-quarter of the rows with T, then one-quarter with F, then one-quarter with T and the last quarter with F. The next column alternates between true and false for each eighth of the rows, then sixteenths, and so on, until the last propositional constant varies between T and F for each row. This will give a complete listing of cases or truth-value assignments possible for those propositional constants.

Argument

The propositional calculus then defines an argument to be a set of propositions. A valid argument is a set of propositions, the last of which follows from—or is implied by—the rest. All other arguments are invalid. The simplest valid argument is modus ponens, one instance of which is the following set of propositions:

1. $P \to Q$
2. P

$\therefore\ Q$

This is a set of three propositions, each line is a proposition, and the last follows from the rest. The first two lines are called premises, and the last line the conclusion. We say that any proposition C follows from any set of propositions $(P_1, \ldots, P_n)$, if C must be true whenever every member of the set $(P_1, \ldots, P_n)$ is true. In the argument above, for any P and Q, whenever P → Q and P are true, necessarily Q is true. Notice that, when P is true, we cannot consider cases 3 and 4 (from the truth table). When P → Q is true, we cannot consider case 2. This leaves only case 1, in which Q is also true. Thus Q is implied by the premises.

This generalizes schematically. Thus, where φ and ψ may be any propositions at all,

1. $\varphi \to \psi$
2. φ

$\therefore\ \psi$

Other argument forms are convenient, but not necessary. Given a complete set of axioms (see below for one such set), modus ponens is sufficient to prove all other argument forms in propositional logic, thus they may be considered to be a derivative. Note, this is not true of the extension of propositional logic to other logics like first-order logic. First-order logic requires at least one additional rule of inference in order to obtain completeness.

The significance of argument in formal logic is that one may obtain new truths from established truths. In the first example above, given the two premises, the truth of Q is not yet known or stated. After the argument is made, Q is deduced. In this way, we define a deduction system to be a set of all propositions that may be deduced from another set of propositions. For instance, given the set of propositions $A = \{P \vee Q, \neg Q \wedge R, (P \vee Q) \to R\}$, we can define a deduction system, Γ, which is the set of all propositions which follow from A. Reiteration is always assumed, so $P \vee Q, \neg Q \wedge R, (P \vee Q) \to R \in \Gamma$. Also, from the first element of A, last element, as well as modus ponens, R is a consequence, and so $R \in \Gamma$. Because we have not included sufficiently complete axioms, though, nothing else may be deduced. Thus, even though most deduction systems studied in propositional logic are able to deduce $(P \vee Q) \leftrightarrow (\neg P \to Q)$, this one is too weak to prove such a proposition.

Generic Description of a Propositional Calculus

A **propositional calculus** is a formal system $\mathcal{L} = \mathcal{L}(\mathrm{A}, \Omega, \mathrm{Z}, \mathrm{I})$, where:

- The alpha set A is a finite set of elements called proposition symbols or propositional variables. Syntactically speaking, these are the most basic elements of the formal language L, otherwise referred to as atomic formulas or terminal elements. In the examples to follow, the elements of A are typically the letters p, q, r, and so on.
- The omega set Ω is a finite set of elements called operator symbols or logical connectives. The set Ω is partitioned into disjoint subsets as follows:

$$\Omega = \Omega_0 \cup \Omega_1 \cup \ldots \cup \Omega_j \cup \ldots \cup \Omega_m.$$

In this partition, Ω_j is the set of operator symbols of arity j.

In the more familiar propositional calculi, Ω is typically partitioned as follows:

$$\Omega_1 = \{\neg\},$$

$$\Omega_2 \subseteq \{\wedge, \vee, \to, \leftrightarrow\}.$$

A frequently adopted convention treats the constant logical values as operators of arity zero, thus:

$$\Omega_0 = \{0, 1\}.$$

Some writers use the tilde (~), or N, instead of ¬; and some use the ampersand (&), the prefixed K, $\cdot$ instead of $\wedge$ Notation varies even more for the set of logical values, with symbols like {false, true}, {F, T}, or $\{\bot, \top\}$ all being seen in various contexts instead of {0, 1}.

- The zeta set Z is a finite set of transformation rules that are called inference rules when they acquire logical applications.
- The iota set I is a finite set of initial points that are called axioms when they receive logical interpretations.

The language of L, also known as its set of formulas, well-formed formulas, is inductively defined by the following rules:

- Base: Any element of the alpha set A is a formula of L.

- If $p_1, p_2, \ldots, p_j$ are formulas and f is in Ω_j, then $(f(p_1, p_2, \ldots, p_j))$ is a formula.
- Closed: Nothing else is a formula of L.

Repeated applications of these rules permits the construction of complex formulas. For example:

1. By rule 1, p is a formula.
2. By rule 2, $\neg p$ is a formula.
3. By rule 1, q is a formula.
4. By rule 2, $(\neg p \lor q)$ is a formula.

Example 1. Simple Axiom System

Let $\mathcal{L}_1 = \mathcal{L}(\mathrm{A}, \Omega, \mathrm{Z}, \mathrm{I})$, where $\mathrm{A}, \Omega, \mathrm{Z}, \mathrm{I}$ are defined as follows:

- The alpha set A, is a finite set of symbols that is large enough to supply the needs of a given discussion, for example:

$\mathrm{A} = \{p, q, r, s, t, u\}$.

- Of the three connectives for conjunction, disjunction, and implication ($\land$, $\lor$, and $\to$), one can be taken as primitive and the other two can be defined in terms of it and negation ($\neg$). Indeed, all of the logical connectives can be defined in terms of a sole sufficient operator. The biconditional ($\leftrightarrow$) can of course be defined in terms of conjunction and implication, with $\leftrightarrow b$ defined as $(a \to b) \land (b \to a)$.

Adopting negation and implication as the two primitive operations of a propositional calculus is tantamount to having the omega set $\Omega = \Omega_1 \cup \Omega_2$ partition as follows:

$\Omega_1 = \{\neg\}$,
$\Omega_2 = \{\to\}$.

- An axiom system discovered by Jan Łukasiewicz formulates a propositional calculus in this language as follows. The axioms are all substitution instances of:

- $(p \to (q \to p))$
- $((p \to (q \to r)) \to ((p \to q) \to (p \to r)))$
- $((\neg p \to \neg q) \to (q \to p))$

- The rule of inference is modus ponens $(p \to q)$, infer q). Then $a \lor b$ is defined as $\neg a \to b$, and $a \land b$ is defined as $\neg(a \to \neg b)$. This system is used in Metamath set.mm formal proof database.

Example 2. Natural Deduction System

Let $\mathcal{L}_2 = \mathcal{L}(\mathrm{A}, \Omega, \mathrm{Z}, \mathrm{I})$, where $\mathrm{A}, \Omega, \mathrm{Z}, \mathrm{I}$ are defined as follows:

- The alpha set A, is a finite set of symbols that is large enough to supply the needs of a given discussion, for example:

$A = \{p, q, r, s, t, u\}$.

- The omega set $\Omega = \Omega_1 \cup \Omega_2$ partitions as follows:

$\Omega_1 = \{\neg\}$,
$\Omega_2 = \{\wedge, \vee, \rightarrow, \leftrightarrow\}$.

In the following example of a propositional calculus, the transformation rules are intended to be interpreted as the inference rules of a so-called natural deduction system. The particular system presented here has no initial points, which means that its interpretation for logical applications derives its theorems from an empty axiom set.

- The set of initial points is empty, that is, $I = \varnothing$.
- The set of transformation rules, Z, is described as follows:

Our propositional calculus has ten inference rules. These rules allow us to derive other true formulas given a set of formulas that are assumed to be true. The first nine simply state that we can infer certain well-formed formulas from other well-formed formulas. The last rule however uses hypothetical reasoning in the sense that in the premise of the rule we temporarily assume an (unproven) hypothesis to be part of the set of inferred formulas to see if we can infer a certain other formula. Since the first nine rules don't do this they are usually described as non-hypothetical rules, and the last one as a hypothetical rule.

In describing the transformation rules, we may introduce a metalanguage symbol $\vdash$. It is basically a convenient shorthand for saying "infer that". The format is $\Gamma \vdash \psi$, in which Γ is a (possibly empty) set of formulas called premises, and ψ is a formula called conclusion. The transformation rule $\Gamma \vdash \psi$ means that if every proposition in Γ is a theorem (or has the same truth value as the axioms), then ψ is also a theorem. Note that considering the following rule Conjunction introduction, we will know whenever Γ has more than one formula, we can always safely reduce it into one formula using conjunction. So for short, from that time on we may represent Γ as one formula instead of a set. Another omission for convenience is when Γ is an empty set, in which case Γ may not appear.

Negation introduction

From $(p \to q)$ and $(p \to \neg q)$, infer $\neg p$.
That is, $\{(p \to q), (p \to \neg q)\} \vdash \neg p$.

Negation elimination

From $\neg p$, infer $(p \to r)$.
That is, $\{\neg p\} \vdash (p \to r)$.

Double negative elimination

From $\neg\neg p$, infer p.
That is, $\neg\neg p \vdash p$.

Conjunction introduction

From p and q, infer $(p \wedge q)$.
That is, $\{p, q\} \vdash (p \wedge q)$.

Conjunction elimination

From $(p \wedge q)$, infer p.
From $(p \wedge q)$, infer q.
That is, $(p \wedge q) \vdash p$ and $(p \wedge q) \vdash q$.

Disjunction introduction

From p, infer $(p \vee q)$.
From q, infer $(p \vee q)$.
That is, $p \vdash (p \vee q)$ and $q \vdash (p \vee q)$

Disjunction elimination

From $(p \vee q)$ and $(p \rightarrow r)$ and $(q \rightarrow r)$, infer r.
That is, $\{p \vee q, p \rightarrow r, q \rightarrow r\} \vdash r$.

Biconditional introduction

From $(p \rightarrow q)$ and $(q \rightarrow p)$, infer $(p \leftrightarrow q)$.
That is, $\{p \rightarrow q, q \rightarrow p\} \vdash (p \leftrightarrow q)$.

Biconditional elimination

From $(p \leftrightarrow q)$, infer $(p \rightarrow q)$.
From $(p \leftrightarrow q)$, infer $(q \rightarrow p)$.
That is, $(p \leftrightarrow q) \vdash (p \rightarrow q)$ and $(p \leftrightarrow q) \vdash (q \rightarrow p)$.

Modus ponens (conditional elimination)

From p and $(p \rightarrow q)$, infer q.
That is, $\{p, p \rightarrow q\} \vdash q$.

Conditional proof (conditional introduction)

From [accepting p allows a proof of q], infer $(p \rightarrow q)$.
That is, $(p \vdash q) \vdash (p \rightarrow q)$.

Basic and Derived Argument Forms

Basic and Derived Argument Forms		
Name	Sequent	Description
Modus Ponens	$((p \to q) \land p) \vdash q$	If p then q; p; therefore q
Modus Tollens	$((p \to q) \land \neg q) \vdash \neg p$	If p then q; not q; therefore not p
Hypothetical Syllogism	$((p \to q) \land (q \to r)) \vdash (p \to r)$	If p then q; if q then r; therefore, if p then r
Disjunctive Syllogism	$((p \lor q) \land \neg p) \vdash q$	Either p or q, or both; not p; therefore, q
Constructive Dilemma	$((p \to q) \land (r \to s) \land (p \lor r)) \vdash (q \lor s)$	If p then q; and if r then s; but p or r; therefore q or s
Destructive Dilemma	$((p \to q) \land (r \to s) \land (\neg q \lor \neg s)) \vdash (\neg p \lor \neg r)$	If p then q; and if r then s; but not q or not s; therefore not p or not r
Bidirectional Dilemma	$((p \to q) \land (r \to s) \land (p \lor \neg s)) \vdash (q \lor \neg r)$	If p then q; and if r then s; but p or not s; therefore q or not r
Simplification	$(p \land q) \vdash p$	p and q are true; therefore p is true
Conjunction	$p, q \vdash (p \land q)$	p and q are true separately; therefore they are true conjointly
Addition	$p \vdash (p \lor q)$	p is true; therefore the disjunction (p or q) is true
Composition	$((p \to q) \land (p \to r)) \vdash (p \to (q \land r))$	If p then q; and if p then r; therefore if p is true then q and r are true
De Morgan's Theorem (1)	$\neg(p \land q) \vdash (\neg p \lor \neg q)$	The negation of (p and q) is equiv. to (not p or not q)
De Morgan's Theorem (2)	$\neg(p \lor q) \vdash (\neg p \land \neg q)$	The negation of (p or q) is equiv. to (not p and not q)
Commutation (1)	$(p \lor q) \vdash (q \lor p)$	(p or q) is equiv. to (q or p)
Commutation (2)	$(p \land q) \vdash (q \land p)$	(p and q) is equiv. to (q and p)
Commutation (3)	$(p \leftrightarrow q) \vdash (q \leftrightarrow p)$	(p is equiv. to q) is equiv. to (q is equiv. to p)
Association (1)	$(p \lor (q \lor r)) \vdash ((p \lor q) \lor r)$	p or (q or r) is equiv. to (p or q) or r
Association (2)	$(p \land (q \land r)) \vdash ((p \land q) \land r)$	p and (q and r) is equiv. to (p and q) and r
Distribution (1)	$(p \land (q \lor r)) \vdash ((p \land q) \lor (p \land r))$	p and (q or r) is equiv. to (p and q) or (p and r)
Distribution (2)	$(p \lor (q \land r)) \vdash ((p \lor q) \land (p \lor r))$	p or (q and r) is equiv. to (p or q) and (p or r)
Double Negation	$p \vdash \neg\neg p$	p is equivalent to the negation of not p
Transposition	$(p \to q) \vdash (\neg q \to \neg p)$	If p then q is equiv. to if not q then not p
Material Implication	$(p \to q) \vdash (\neg p \lor q)$	If p then q is equiv. to not p or q
Material Equivalence (1)	$(p \leftrightarrow q) \vdash ((p \to q) \land (q \to p))$	(p iff q) is equiv. to (if p is true then q is true) and (if q is true then p is true)
Material Equivalence (2)	$(p \leftrightarrow q) \vdash ((p \land q) \lor (\neg p \land \neg q))$	(p iff q) is equiv. to either (p and q are true) or (both p and q are false)
Material Equivalence (3)	$(p \leftrightarrow q) \vdash ((p \lor \neg q) \land (\neg p \lor q))$	(p iff q) is equiv to., both (p or not q is true) and (not p or q is true)
Exportation[9]	$((p \land q) \to r) \vdash (p \to (q \to r))$	from (if p and q are true then r is true) we can prove (if q is true then r is true, if p is true)
Importation	$(p \to (q \to r)) \vdash ((p \land q) \to r)$	If p then (if q then r) is equivalent to if p and q then r
Tautology (1)	$p \vdash (p \lor p)$	p is true is equiv. to p is true or p is true
Tautology (2)	$p \vdash (p \land p)$	p is true is equiv. to p is true and p is true
Tertium non datur (Law of Excluded Middle)	$\vdash (p \lor \neg p)$	p or not p is true
Law of Non-Contradiction	$\vdash \neg(p \land \neg p)$	p and not p is false, is a true statement

Proofs in Propositional Calculus

One of the main uses of a propositional calculus, when interpreted for logical applications, is to determine relations of logical equivalence between propositional formulas. These relationships are determined by means of the available transformation rules, sequences of which are called derivations or proofs.

In the discussion to follow, a proof is presented as a sequence of numbered lines, with each line consisting of a single formula followed by a reason or justification for introducing that formula. Each premise of the argument, that is, an assumption introduced as an hypothesis of the argument, is listed at the beginning of the sequence and is marked as a "premise" in lieu of other justification. The conclusion is listed on the last line. A proof is complete if every line follows from the previous ones by the correct application of a transformation rule. (For a contrasting approach, see proof-trees).

Example of a Proof

- To be shown that A → A.

Example of a Proof		
Number	Formula	Reason
1	A	premise
2	$A \lor A$	From (1) by disjunction introduction
3	$(A \lor A) \land A$	From (1) and (2) by conjunction introduction
4	A	From (3) by conjunction elimination
5	$A \vdash A$	Summary of (1) through (4)
6	$\vdash A \to A$	From (5) by conditional proof

Interpret $A \vdash A$ as "Assuming A, infer A". Read $\vdash A \to A$ as "Assuming nothing, infer that A implies A", or "It is a tautology that A implies A", or "It is always true that A implies A".

Soundness and Completeness of the Rules

The crucial properties of this set of rules are that they are sound and complete. Informally this means that the rules are correct and that no other rules are required. These claims can be made more formal as follows.

We define a truth assignment as a function that maps propositional variables to **true** or **false**. Informally such a truth assignment can be understood as the description of a possible state of affairs (or possible world) where certain statements are true and others are not. The semantics of formulas can then be formalized by defining for which "state of affairs" they are considered to be true, which is what is done by the following definition.

We define when such a truth assignment A satisfies a certain well-formed formula with the following rules:

- A satisfies the propositional variable P if and only if A(P) = true

- A satisfies $\neg\varphi$ if and only if A does not satisfy φ
- A satisfies $(\varphi \wedge \psi)$ if and only if A satisfies both φ and ψ
- A satisfies $(\varphi \vee \psi)$ if and only if A satisfies at least one of either φ or ψ
- A satisfics $(\varphi \rightarrow \psi)$ if and only if it is not the case that A satisfies φ but not ψ
- A satisfies $(\varphi \leftrightarrow \psi)$ if and only if A satisfies both φ and ψ or satisfies neither one of them

With this definition we can now formalize what it means for a formula φ to be implied by a certain set S of formulas. Informally this is true if in all worlds that are possible given the set of formulas S the formula φalso holds. This leads to the following formal definition: We say that a set S of well-formed formulas semantically entails (or implies) a certain well-formed formula φ if all truth assignments that satisfy all the formulas in S also satisfy φ.

Finally we define syntactical entailment such that φ is syntactically entailed by S if and only if we can derive it with the inference rules that were presented above in a finite number of steps. This allows us to formulate exactly what it means for the set of inference rules to be sound and complete:

Soundness: If the set of well-formed formulas S syntactically entails the well-formed formula φ then Ssemantically entails φ.

Completeness: If the set of well-formed formulas S semantically entails the well-formed formula φ then S syntactically entails φ.

For the above set of rules this is indeed the case.

Sketch of a Soundness Proof

(For most logical systems, this is the comparatively "simple" direction of proof)

Notational conventions: Let G be a variable ranging over sets of sentences. Let A, B and C range over sentences. For "G syntactically entails A" we write "G proves A". For "G semantically entails A" we write "G implies A".

We want to show: (A)(G) (if G proves A, then G implies A).

We note that "G proves A" has an inductive definition, and that gives us the immediate resources for demonstrating claims of the form "If G proves A, then ...". So our proof proceeds by induction.

- Basis. Show: If A is a member of G, then G implies A.
- Basis. Show: If A is an axiom, then G implies A.
- Inductive step (induction on n, the length of the proof):
    - Assume for arbitrary G and A that if G proves A in n or fewer steps, then G implies A.
    - For each possible application of a rule of inference at step n + 1, leading to a new theorem B, show that G implies B.

Notice that Basis Step II can be omitted for natural deduction systems because they have no axioms. When used, Step II involves showing that each of the axioms is a (semantic) logical truth. The Basis steps demonstrate that the simplest provable sentences from G are also implied by G, for any G. (The proof is simple, since the semantic fact that a set implies any of its members, is also trivial.) The Inductive step will systematically cover all the further sentences that might be provable—by considering each case where we might reach a logical conclusion using an inference rule—and shows that if a new sentence is provable, it is also logically implied. (For example, we might have a rule telling us that from "A" we can derive "A or B". In III.a We assume that if A is provable it is implied. We also know that if A is provable then "A or B" is provable. We have to show that then "A or B" too is implied. We do so by appeal to the semantic definition and the assumption we just made. A is provable from G, we assume. So it is also implied by G. So any semantic valuation making all of G true makes A true. But any valuation making A true makes "A or B" true, by the defined semantics for "or". So any valuation which makes all of G true makes "A or B" true. So "A or B" is implied.) Generally, the Inductive step will consist of a lengthy but simple case-by-case analysis of all the rules of inference, showing that each "preserves" semantic implication.

By the definition of provability, there are no sentences provable other than by being a member of G, an axiom, or following by a rule; so if all of those are semantically implied, the deduction calculus is sound.

Sketch of Completeness Proof

(This is usually the much harder direction of proof.)

We adopt the same notational conventions as above.

We want to show: If G implies A, then G proves A. We proceed by contraposition: We show instead that if G does **not** prove A then G does **not** imply A.

- G does not prove A. (Assumption)
- If G does not prove A, then we can construct an (infinite) **Maximal Set**, G^*, which is a superset of G and which also does not prove A.
  - Place an ordering on all the sentences in the language (e.g., shortest first, and equally long ones in extended alphabetical ordering), and number them (E_1, E_2, ...)
  - Define a series G_n of sets (G_0, G_1, ...) inductively:

i. $G_0 = G$

ii. If $G_k \cup \{E_{k+1}\}$ proves A, then $G_{k+1} = G_k$

iii. If $G_k \cup \{E_{k+1}\}$ does not prove A, then $G_{k+1} = G_k \cup \{E_{k+1}\}$

  - Define G^* as the union of all the G_n. (That is, G^* is the set of all the sentences that are in any G_n.)
  - It can be easily shown that
    i. G^* contains (is a superset of) G (by (b.i));
    ii. G^* does not prove A (because if it proves A then some sentence was added to some G_n which caused it to prove A; but this was ruled out by definition); and

iii. G^* is a Maximal Set with respect to A: If any more sentences whatever were added to G^*, it would prove A. (Because if it were possible to add any more sentences, they should have been added when they were encountered during the construction of the G_n, again by definition)

- If G^* is a Maximal Set with respect to A, then it is **truth-like**. This means that it contains C only if it does **not** contain ¬C; If it contains C and contains "If C then B" then it also contains B; and so forth.
- If G^* is truth-like there is a G^*-Canonical valuation of the language: one that makes every sentence in G^* true and everything outside G^* false while still obeying the laws of semantic composition in the language.
- A G^*-canonical valuation will make our original set G all true, and make A false.
- If there is a valuation on which G are true and A is false, then G does not (semantically) imply A.

QED

Another Outline for a Completeness Proof

If a formula is a tautology, then there is a truth table for it which shows that each valuation yields the value true for the formula. Consider such a valuation. By mathematical induction on the length of the subformulas, show that the truth or falsity of the subformula follows from the truth or falsity (as appropriate for the valuation) of each propositional variable in the subformula. Then combine the lines of the truth table together two at a time by using "(P is true implies S) implies ((P is false implies S) implies S)". Keep repeating this until all dependencies on propositional variables have been eliminated. The result is that we have proved the given tautology. Since every tautology is provable, the logic is complete.

Interpretation of a truth-Functional Propositional Calculus

An **interpretation of a truth-functional propositional calculus** P is an assignment to each propositional symbol of P of one or the other (but not both) of the truth values truth (**T**) and falsity (**F**), and an assignment to the connective symbols of Pof their usual truth-functional meanings. An interpretation of a truth-functional propositional calculus may also be expressed in terms of truth tables.

For n distinct propositional symbols there are 2^n distinct possible interpretations. For any particular symbol a, for example, there are $2^1 = 2$ possible interpretations:

- A is assigned **T**, or
- A is assigned **F**.

For the pair a, b there are $2^2 = 4$ possible interpretations:

- both are assigned **T**,
- both are assigned **F**,
- a is assigned **T** and b is assigned **F**, or
- a is assigned **F** and b is assigned **T**.

Since $\mathcal{P}$ has $\aleph_0$, that is, denumerably many propositional symbols, there are $2^{\aleph_0} = c$, and therefore uncountably many distinct possible interpretations of $\mathcal{P}$.

Interpretation of a Sentence of Truth-Functional Propositional Logic

If φ and ψ are formulas of $\mathcal{P}$ and $\mathcal{I}$ is an interpretation of $\mathcal{P}$ then:

- A sentence of propositional logic is true under an interpretation $\mathcal{I}$ iff $\mathcal{I}$ assigns the truth value **T** to that sentence. If a sentence is true under an interpretation, then that interpretation is called a model of that sentence.
- φ is false under an interpretation $\mathcal{I}$ iff φ is not true under $\mathcal{I}$.
- A sentence of propositional logic is logically valid if it is true under every interpretation

$\models \phi$ means that φ is logically valid

- A sentence ψ of propositional logic is a semantic consequence of a sentence φ iff there is no interpretation under which φ is true and ψ is false.
- A sentence of propositional logic is consistent iff it is true under at least one interpretation. It is inconsistent if it is not consistent.

Some consequences of these definitions:

- For any given interpretation a given formula is either true or false.
- No formula is both true and false under the same interpretation.
- φ is false for a given interpretation iff $\neg\phi$ is true for that interpretation; and φ is true under an interpretation iff $\neg\phi$ is false under that interpretation.
- If φ and $(\phi \to \psi)$ are both true under a given interpretation, then ψ is true under that interpretation.
- If $\models_{\mathrm{P}} \phi$ and $\models_{\mathrm{P}} (\phi \to \psi)$, then $\models_{\mathrm{P}} \psi$..
- $\neg\phi$ is true under $\mathcal{I}$ iff φ is not true under $\mathcal{I}$..
- $(\phi \to \psi)$ is true under $\mathcal{I}$ iff either φ is not true under $\mathcal{I}$ or ψ is true under $\mathcal{I}$..
- A sentence ψ of propositional logic is a semantic consequence of a sentence φ iff φ iff $(\phi \to \psi)$ is logically valid, that is $\phi \models_{\mathrm{P}} \psi$ iff $\models_{\mathrm{P}} (\phi \to \psi)$.

Alternative Calculus

It is possible to define another version of propositional calculus, which defines most of the syntax of the logical operators by means of axioms, and which uses only one inference rule.

Axioms

Let φ, χ, and ψ stand for well-formed formulas. (The well-formed formulas themselves would not contain any Greek letters, but only capital Roman letters, connective operators, and parentheses.) Then the axioms are as follows:

- Axiom THEN-2 may be considered to be a "distributive property of implication with respect to implication."
- Axioms AND-1 and AND-2 correspond to "conjunction elimination". The relation between AND-1 and AND-2 reflects the commutativity of the conjunction operator.
- Axiom AND-3 corresponds to "conjunction introduction."
- Axioms OR-1 and OR-2 correspond to "disjunction introduction." The relation between OR-1 and OR-2 reflects the commutativity of the disjunction operator.
- Axiom NOT-1 corresponds to "reductio ad absurdum."
- Axiom NOT-2 says that "anything can be deduced from a contradiction."
- Axiom NOT-3 is called "tertium non datur" (Latin: "a third is not given") and reflects the semantic valuation of propositional formulas: a formula can have a truth-value of either true or false. There is no third truth-value, at least not in classical logic. Intuitionistic logicians do not accept the axiom NOT-3.

Inference rule

The inference rule is modus ponens:

$\phi, \phi \to \chi \vdash \chi.$

Meta-inference Rule

Let a demonstration be represented by a sequence, with hypotheses to the left of the turnstile and the conclusion to the right of the turnstile. Then the deduction theorem can be stated as follows:

If the sequence

$$\phi_1, \phi_2, \ldots, \phi_n, \chi \vdash \psi$$

has been demonstrated, then it is also possible to demonstrate the sequence

$$\phi_1, \phi_2, \ldots, \phi_n \vdash \chi \to \psi.$$

This deduction theorem (DT) is not itself formulated with propositional calculus: it is not a theorem of propositional calculus, but a theorem about propositional calculus.

In this sense, it is a meta-theorem, comparable to theorems about the soundness or completeness of propositional calculus.

On the other hand, DT is so useful for simplifying the syntactical proof process that it can be considered and used as another inference rule, accompanying modus ponens. In this sense, DT corresponds to the natural conditional proof inference rule which is part of the first version of propositional calculus introduced in this article.

The converse of DT is also valid:

If the sequence

$\phi_1, \phi_2, \ldots, \phi_n \vdash \chi \to \psi$

has been demonstrated, then it is also possible to demonstrate the sequence

$\phi_1, \phi_2, \ldots, \phi_n, \chi \vdash \psi$

in fact, the validity of the converse of DT is almost trivial compared to that of DT:

If

$\phi_1, \ldots, \phi_n \vdash \chi \to \psi$

then

1: $\phi_1, \ldots, \phi_n, \chi \vdash \chi \to \psi$
2: $\phi_1, \ldots, \phi_n, \chi \vdash \chi$

and from (1) and (2) can be deduced

3: $\phi_1, \ldots, \phi_n, \chi \vdash \psi$

by means of modus ponens, Q.E.D.

The converse of DT has powerful implications: it can be used to convert an axiom into an inference rule. For example, the axiom AND-1,

$\vdash \phi \wedge \chi \to \phi$

can be transformed by means of the converse of the deduction theorem into the inference rule

$\phi \wedge \chi \vdash \phi$

which is conjunction elimination, one of the ten inference rules used in the first version (in this article) of the propositional calculus.

Example of a proof

The following is an example of a (syntactical) demonstration, involving only axioms THEN-1 and THEN-2:

Prove: $A \to A$ (Reflexivity of implication).

Proof:

1. $(A \to ((B \to A) \to A)) \to ((A \to (B \to A)) \to (A \to A))$

   Axiom THEN-2 with $\phi = A, \chi = B \to A, \psi = A$

2. $A \to ((B \to A) \to A)$

   Axiom THEN-1 with $\phi = A, \chi = B \to A$

3. $(A \to (B \to A)) \to (A \to A)$

   From (1) and (2) by modus ponens.

4. $A \to (B \to A)$

   Axiom THEN-1 with $\phi = A, \chi = B$

5. $A \to A$

   From (3) and (4) by modus ponens.

Equivalence to Equational Logics

The preceding alternative calculus is an example of a Hilbert-style deduction system. In the case of propositional systems the axioms are terms built with logical connectives and the only inference rule is modus ponens. Equational logic as standardly used informally in high school algebra is a different kind of calculus from Hilbert systems. Its theorems are equations and its inference rules express the properties of equality, namely that it is a congruence on terms that admits substitution.

Classical propositional calculus as described above is equivalent to Boolean algebra, while intuitionistic propositional calculus is equivalent to Heyting algebra. The equivalence is shown by translation in each direction of the theorems of the respective systems. Theorems ϕ of classical or intuitionistic propositional calculus are translated as equations $\phi = 1$ of Boolean or Heyting algebra respectively. Conversely theorems $x = y$ of Boolean or Heyting algebra are translated as theorems ϕ of classical or intuitionistic calculus respectively, for which $x \equiv y$ is a standard abbreviation. In the case of Boolean algebra $x = y$ can also be translated $(x \wedge y) \vee (\neg x \wedge \neg y)$, but this translation is incorrect intuitionistically.

In both Boolean and Heyting algebra, inequality $x \leq y$ can be used in place of equality. The equality $x = y$ is expressible as a pair of inequalities $x \leq y$ and $y \leq x$. Conversely the inequality $x \leq y$ is expressible as the equality $x \wedge y = x$, or as $x \vee y = y$. The significance of inequality for Hilbert-style systems is that it corresponds to the latter's deduction or entailment symbol $\vdash$. An entailment

$$\phi_1, \phi_2, \ldots, \phi_n \vdash \psi$$

is translated in the inequality version of the algebraic framework as

$$\phi_1 \wedge \phi_2 \wedge \ldots \wedge \phi_n \leq \psi$$

Conversely the algebraic inequality $x \leq y$ is translated as the entailment

$$x \vdash y.$$

The difference between implication $x \to y$ and inequality or entailment $x \leq y$ or $x \vdash y$ is that the former is internal to the logic while the latter is external. Internal implication between two terms is another term of the same kind. Entailment as external implication between two terms expresses a metatruth outside the language of the logic, and is considered part of the metalanguage. Even when the logic under study is intuitionistic, entailment is ordinarily understood classically as two-valued: either the left side entails, or is less-or-equal to, the right side, or it is not.

Similar but more complex translations to and from algebraic logics are possible for natural deduction systems as described above and for the sequent calculus. The entailments of the latter can be interpreted as two-valued, but a more insightful interpretation is as a set, the elements of which can be understood as abstract proofs organized as the morphisms of a category. In this interpretation the cut rule of the sequent calculus corresponds to composition in the category. Boolean and Heyting algebras enter this picture as special categories having at most one morphism per homset, i.e., one proof per entailment, corresponding to the idea that existence of proofs is all that matters: any proof will do and there is no point in distinguishing them.

Graphical Calculi

It is possible to generalize the definition of a formal language from a set of finite sequences over a finite basis to include many other sets of mathematical structures, so long as they are built up by finitary means from finite materials. What's more, many of these families of formal structures are especially well-suited for use in logic.

For example, there are many families of graphs that are close enough analogues of formal languages that the concept of a calculus is quite easily and naturally extended to them. Indeed, many species of graphs arise as parse graphs in the syntactic analysis of the corresponding families of text structures. The exigencies of practical computation on formal languages frequently demand that text strings be converted into pointer structure renditions of parse graphs, simply as a matter of checking whether strings are well-formed formulas or not. Once this is done, there are many advantages to be gained from developing the graphical analogue of the calculus on strings. The mapping from strings to parse graphs is called parsing and the inverse mapping from parse graphs to strings is achieved by an operation that is called traversing the graph.

Other Logical Calculi

Propositional calculus is about the simplest kind of logical calculus in current use. It can be extended in several ways. (Aristotelian "syllogistic" calculus, which is largely supplanted in modern logic, is in some ways simpler – but in other ways more complex – than propositional calculus.) The most immediate way to develop a more complex logical calculus is to introduce rules that are sensitive to more fine-grained details of the sentences being used.

First-order logic (a.k.a. first-order predicate logic) results when the "atomic sentences" of propositional logic are broken up into terms, variables, predicates, and quantifiers, all keeping the rules of propositional logic with some new ones introduced. (For example, from "All dogs are mammals" we may infer "If Rover is a dog then Rover is a mammal".) With the tools of first-order logic it is possible to formulate a number of theories, either with explicit axioms or by rules of inference, that can themselves be treated as logical calculi. Arithmetic is the best known of these; others include set theory and mereology. Second-order logic and other higher-order logics are formal extensions of first-order logic. Thus, it makes sense to refer to propositional logic as "zeroth-order logic", when comparing it with these logics.

Modal logic also offers a variety of inferences that cannot be captured in propositional calculus. For example, from "Necessarily p" we may infer that p. From p we may infer "It is possible that p". The translation between modal logics and algebraic logics concerns classical and intuitionistic logics but with the introduction of a unary operator on Boolean or Heyting algebras, different from the Boolean operations, interpreting the possibility modality, and in the case of Heyting algebra a second operator interpreting necessity (for Boolean algebra this is redundant since necessity is the De Morgan dual of possibility). The first operator preserves 0 and disjunction while the second preserves 1 and conjunction.

Many-valued logics are those allowing sentences to have values other than true and false. (For example, neither and both are standard "extra values"; "continuum logic" allows each sentence to have any of an infinite number of "degrees of truth" between true and false.) These logics often require calculational devices quite distinct from propositional calculus. When the values form a Boolean algebra (which may have more than two or even infinitely many values), many-valued logic reduces to classical logic; many-valued logics are therefore only of independent interest when the values form an algebra that is not Boolean.

SORITES PARADOX

The sorites paradox is a paradox that arises from vague predicates. A typical formulation involves a heap of sand, from which grains are individually removed. Under the assumption that removing a single grain does not turn a heap into a non-heap, the paradox is to consider what happens when the process is repeated enough times: is a single remaining grain still a heap? If not, when did it change from a heap to a non-heap?

The sorites paradox: If a heap is reduced by a single grain at a time, at what exact point does it cease to be considered a heap?

The Original Formulation and Variations

Paradox of the Heap

The word "sorites" derives from the Greek word for heap. The paradox is so named because of its original characterization, attributed to Eubulides of Miletus. The paradox goes as follows: consider a heap of sand from which grains are individually removed. One might construct the argument, using premises, as follows:

1000000 grains of sand is a heap of sand (Premise 1)

A heap of sand minus one grain is still a heap. (Premise 2)

Repeated applications of Premise 2 (each time starting with one fewer grain) eventually forces one to accept the conclusion that a heap may be composed of just one grain of sand.). Read (1995) observes that "the argument is itself a heap, or sorites, of steps of modus ponens":

1000000 grains is a heap.

If 1000000 grains is a heap then 999999 grains is a heap.

So 999999 grains is a heap.

If 999999 grains is a heap then 999998 grains is a heap.

So 999998 grains is a heap.

If ...

... So 1 grain is a heap.

Variations

Color gradient illustrating a sorites paradox, any adjacent colors being indistinguishable by the human eye

Another formulation is to start with a grain of sand, which is clearly not a heap, and then assume that adding a single grain of sand to something that is not a heap does not turn it into a heap. Inductively, this process can be repeated as much as one wants without ever constructing a heap. A more natural formulation of this variant is to assume a set of colored chips exists such that two adjacent chips vary in color too little for human eyesight to be able to distinguish between them. Then by induction on this premise, humans would not be able to distinguish between any colors.

The removal of one drop from the ocean, will not make it 'not an ocean' (it is still an ocean), but since the volume of water in the ocean is finite, eventually, after enough removals, even a litre of water left is still an ocean.

This paradox can be reconstructed for a variety of predicates, for example, with "tall", "rich", "old", "blue", "bald", and so on. Bertrand Russell argued that all of natural language, even logical connectives, is vague; moreover, representations of propositions are vague.

Other similar paradoxes are:

- Argument of the beard
- The bald man paradox

Proposed Resolutions

On the face of it, there are some ways to avoid this conclusion. One may object to the first premise by denying 1000000grains of sand makes a heap. But 1000000 is just an arbitrarily large number, and the argument will go through with any such number. So the response must deny outright that there are such things as heaps. Peter Unger defends this solution.Alternatively, one may object to the second premise by stating that it is not true for all heaps of sand that removing one grain from it still makes a heap.

Setting a Fixed Boundary

A common first response to the paradox is to call any set of grains that has more than a certain number of grains in it a heap. If one were to set the "fixed boundary" at, say, 10000 grains then one would claim that for fewer than 10000, it is not a heap; for 10000 or more, then it is a heap.

However, such solutions are unsatisfactory as there seems little significance to the difference between 9999 grains and 10000grains. The boundary, wherever it may be set, remains as arbitrary and so its precision is misleading. It is objectionable on both philosophical and linguistic grounds: the former on account of its arbitrariness, and the latter on the ground that it is simply not how we use natural language.

Unknowable Boundaries (or Epistemicism)

Timothy Williamson and Roy Sorensen hold an approach that there are fixed boundaries but that they are necessarily unknowable.

Supervaluationism

Supervaluationism is a semantics for dealing with irreferential singular terms and vagueness. It allows one to retain the usual tautological laws even when dealing with undefined truth values. As an example for a proposition about an irreferential singular term, consider the sentence "Pegasus likes licorice". Since the name "Pegasus" fails to refer, no truth value can be assigned to the sentence; there is nothing in the myth that would justify any such assignment. However, there are some statements about "Pegasus" which have definite truth values nevertheless, such as "Pegasus likes licorice or Pegasus doesn't like licorice". This sentence is an instance of the tautology "$p \vee \neg p$"", i.e. the valid schema "*p or not-p*". According to supervaluationism, it should be true regardless of whether or not its components have a truth value.

Similarly, "1000 grains of sand is a heap of sand" may be considered a border case having no truth value, but "1000 grains of sand is a heap of sand, or 1000 grains of sand is not a heap of sand" should be true.

Precisely, let v be a classical valuation defined on every atomic sentence of the language L, and let At(x) be the number of distinct atomic sentences in x. Then for every sentence x, at most $2^{At(x)}$ distinct classical valuations can exist. A supervaluation vis a function from sentences to truth values such that, a sentence {\displaystyle x} is super-true $V(x) = \text{True})$ if and only if $v(x) = \text{True}$ for every classical valuation v; likewise for super-false. Otherwise, $V(x)$ is undefined—i.e. exactly when there are two classical valuations v and v' such that $v(x) = \text{True}$ and $v'(x) = \text{False}$.

For example, let $L p$ be the formal translation of "Pegasus likes licorice". Then there are exactly two classical valuations v and v' on $L p$, viz. $v(L p) = \text{True}$ and $v'(L p) = \text{False}$. So $L p$ is neither super-true nor super-false. However, the tautology $L p \vee \neg L p$ is evaluated to True by every classical valuation; it is hence super-true. Similarly, the formalization of the above heap proposition $H 1000$ is neither super-true nor super-false, but $H 1000 \vee \neg H 1000$ is super-true.

Truth Gaps, Gluts, and Many-Valued Logics

Another approach is to use a multi-valued logic. From this point of view, the problem is with the principle of bivalence: the sand is either a heap or is not a heap, without any shades of gray. Instead of two logical states, heap and not-heap, a three value system can be used, for example heap, indeterminate and not-heap. However, three valued systems do not truly resolve the paradox as there is still a dividing line between heap and indeterminate and also between indeterminate and not-heap. The third truth-value can be understood either as a truth-value gap or as a truth-value glut.

Alternatively, fuzzy logic offers a continuous spectrum of logical states represented in the unit interval of real numbers [0,1]—it is a many-valued logic with infinitely-many truth-values, and thus the sand moves smoothly from "definitely heap" to "definitely not heap", with shades in the intermediate region. Fuzzy hedges are used to divide the continuum into regions corresponding to classes like definitely heap, mostly heap, partly heap, slightly heap, and not heap.

Hysteresis

Another approach, introduced by Raffman, is to use hysteresis, that is, knowledge of what the collection of sand started as. Equivalent amounts of sand may be called heaps or not based on how they got there. If a large heap (indisputably described as a heap) is slowly diminished, it preserves its "heap status" to a point, even as the actual amount of sand is reduced to a smaller number of grains. For example, suppose 500 grains is a pile and 1000 grains is a heap. There will be an overlap for these states. So if one is reducing it from a heap to a pile, it is a heap going down until, say, 750. At that point one would stop calling it a heap and start calling it a pile. But if one replaces one grain, it would not instantly turn back into a heap. When going up it would remain a pile until, say, 900 grains. The numbers picked are arbitrary; the point is, that the same amount can be either a heap or a pile depending on what it was before the change. A common use of hysteresis would be the thermostat for air conditioning: the AC is set at 77 °F and it then cools down to just below 77 °F, but does not turn on again instantly at 77.001

°F—it waits until almost 78 °F, to prevent instant change of state over and over again.

Group Consensus

One can establish the meaning of the word "heap" by appealing to consensus. This approach claims that a collection of grains is as much a "heap" as the proportion of people in a group who believe it to be so. In other words, the probability that any collection is considered a heap is the expected value of the distribution of the group's views.

A group may decide that:

- One grain of sand on its own is not a heap.
- A large collection of grains of sand is a heap.

Between the two extremes, individual members of the group may disagree with each other over whether any particular collection can be labelled a "heap". The collection can then not be definitively claimed to be a "heap" or "not a heap". This can be considered an appeal to descriptive linguistics rather than prescriptive linguistics, as it resolves the issue of definition based on how the population uses natural language. Indeed, if a precise prescriptive definition of "heap" is available then the group consensus will always be unanimous and the paradox does not arise.

Dropping transitivity of the relations involved

In the above color example, the argument is tacitly based on considering the relation "for the human eye, color X is indistinguishable from Y" as an equivalence relation, in particular as transitive. To drop the transitivity assumption is a possibility to resolve the paradox.

Modelling "*X* more or equally red than *Y*" as quasitransitive (Q) and as transitive (T) relation						
*x**Y*	f01000	e02000	d03000	c04000	b05000	a06000
f01000	QT	QT	QTP	QTP	QTP	QTP
e02000	Q	QT	QT	QTP	QTP	QTP
d03000		Q	QT	QT	QTP	QTP
c04000			Q	QT	QT	QTP
b05000				Q	QT	QT
a06000					Q	QT

Similarly, the paradox is based on considering the relation "for the human eye, color X looks more or equally red than Y" as a reflexive total ordering; again, dropping its transitivity resolves the paradox.

Instead, the relation between colors can be described as a quasitransitive relation, employing a concept introduced by microeconomist Amartya Sen in 1969. The table shows a simple example, with color differences overdone for readability. A "Q" and a "T" indicates that the row's color looks more or equally red than column's color in the quasitransitive and the transitive version of the relation, respectively. In the quasitransitive version, e.g. the colors f01000 and e02000 are modelled as indistinguishable, since a "Q" appears in both their intersection cells. A "P" indicates the asymmetric part of the quasitransitive version.

To resolve the original heap variation of the paradox with this approach, the relation "X grains are more a heap than Y grains" should be considered quasitransitive rather than transitive.

REFERENCES

1. Bobzien, Susanne (1 January 2016). Zalta, Edward N., ed. The Stanford Encyclopedia of Philosophy – via Stanford Encyclopedia of Philosophy.
2. Marenbon, John (2007). Medieval philosophy: an historical and philosophical introduction. Routledge. p. 137.
3. Peckhaus, Volker (1 January 2014). Zalta, Edward N., ed. The Stanford Encyclopedia of Philosophy – via Stanford Encyclopedia of Philosophy.
4. Hurley, Patrick (2007). A Concise Introduction to Logic 10th edition. Wadsworth Publishing. p. 392.
5. Beth, Evert W.; "Semantic entailment and formal derivability", series: Mededlingen van de Koninklijke Nederlandse Akademie van Wetenschappen, Afdeling Letterkunde, Nieuwe Reeks, vol. 18, no. 13, Noord-Hollandsche Uitg. Mij., Amsterdam, 1955, pp. 309–42. Reprinted in Jaakko Intikka (ed.) The Philosophy of Mathematics, Oxford University Press, 1969
6. Truth in Frege
7. "Russell: the Journal of Bertrand Russell Studies".
8. Wernick, William (1942) "Complete Sets of Logical Functions," Transactions of the American Mathematical Society **51**, pp. 117–132.
9. Toida, Shunichi (2 August 2009). "Proof of Implications". CS381 Discrete Structures/ Discrete Mathematics Web Course Material. Department Of Computer Science, Old Dominion University. Retrieved 10 March 2010.
10. Hunter, Geoffrey (1971). Metalogic: An Introduction to the Metatheory of Standard First-Order Logic. University of California Pres. ISBN 0-520-02356-0.
11. Lou Goble (2001). The Blackwell guide to philosophical logic. Wiley-Blackwell. p. 309. ISBN 978-0-631-20693-4.
12. Paul Tomassi (1999). Logic. Routledge. p. 124. ISBN 978-0-415-16696-6.
13. Lou Goble (2001). The Blackwell guide to philosophical logic. Wiley-Blackwell. p. 4. ISBN 978-0-631-20693-4.
14. Mark Hürlimann (2009). Dealing with Real-World Complexity: Limits, Enhancements and New Approaches for Policy Makers. Gabler Verlag. p. 42. ISBN 978-3-8349-1493-4.
15. Dov M. Gabbay; John Woods (2007). The Many Valued and Nonmonotonic Turn in Logic. The handbook of the history of logic. **8**. Elsevier. p. vii. ISBN 978-0-444-51623-7. Retrieved 4 April 2011.
16. Graham Priest (2008). An introduction to non-classical logic: from if to is. Cambridge University Press. pp. 124–125. ISBN 978-0-521-85433-7.
17. Morten Heine Sørensen; Paweł Urzyczyn (2006). Lectures on the Curry-Howard isomorphism. Elsevier. pp. 206–207. ISBN 978-0-444-52077-7.
18. Shramko, Y.; Wansing, H. (2015). "Truth Values, Stanford Encyclopedia of Philosophy".

19. Note the use of the (extremely) definite article: " This " as opposed to a more-vague " The ". " The " would have to be accompanied with a pointing-gesture to make it definitive. Ff Principia Mathematica (2nd edition), p. 91. Russell & Whitehead observe that this " this " indicates "something given in sensation" and as such it shall be considered "elementary".

20. Stephen C. Kleene 1952 Introduction to Metamathematics, 6th Reprint 1971, North-Holland Publishing Company, Amsterdam NY, ISBN 0-7294-2130-9.

21. "Strong tables" is Kleene's choice of words. Note that even though " u " may appear for the value of Q or R, " t " or " f " may, in those occasions, appear as a value in " Q V R ", " Q & R " and " Q → R ". "Weak tables" on the other hand, are "regular", meaning they have " u " appear in all cases when the value " u " is applied to either Q or R or both. Kleene notes that these tables are not the same as the original values of the tables of Łukasiewicz 1920. (Kleene gives these differences on page 335). He also concludes that " u " can mean any or all of the following: "undefined", "unknown (or value immaterial)", "value disregarded for the moment", i.e. it is a third category that does not (ultimately) exclude " t " and " f " (page 335).

22. Copi and Cohen

23. Hurley, Patrick J. (2015), A Concise Introduction to Logic (12th ed.), Cengage Learning, ISBN 978-1-285-19654-1

24. Moore and Parker

25. Boolean Algebra by R. L. Goodstein. ISBN 0-486-45894-6

26. 2000 Solved Problems in Digital Electronics by S. P. Bali

27. DeMorgan's Theorems at mtsu.edu

28. Bocheński's History of Formal Logic

29. William of Ockham, Summa Logicae, part II, sections 32 and 33.

30. Jean Buridan, Summula de Dialectica. Trans. Gyula Klima. New Haven: Yale University Press, 2001. See especially Treatise 1, Chapter 7, Section 5. ISBN 0-300-08425-0

31. Augustus De Morgan (1806–1871) by Robert H. Orr

32. Andrews, Peter B. (2002); An Introduction to Mathematical Logic and Type Theory: To Truth Through Proof, 2nd ed., Berlin: Kluwer Academic Publishers. Available from Springer.

33. Avigad, Jeremy; Donnelly, Kevin; Gray, David; and Raff, Paul (2007); "A formally verified proof of the prime number theorem", ACM Transactions on Computational Logic, vol. 9 no. 1 doi:10.1145/1297658.1297660

34. Barwise, Jon (1977); "An Introduction to First-Order Logic", in Barwise, Jon, ed. (1982). Handbook of Mathematical Logic. Studies in Logic and the Foundations of Mathematics. Amsterdam, NL: North-Holland. ISBN 978-0-444-86388-1.

35. Barwise, Jon; and Etchemendy, John (2000); Language Proof and Logic, Stanford, CA: CSLI Publications (Distributed by the University of Chicago Press)

36. Bocheński, Józef Maria (2007); A Précis of Mathematical Logic, Dordrecht, NL: D. Reidel, translated from the French and German editions by Otto Bird

37. Ferreirós, José (2001); The Road to Modern Logic — An Interpretation, Bulletin of

Symbolic Logic, Volume 7, Issue 4, 2001, pp. 441–484, DOI 10.2307/2687794, JStor

38. Gamut, L. T. F. (1991); Logic, Language, and Meaning, Volume 2: Intensional Logic and Logical Grammar, Chicago, IL: University of Chicago Press, ISBN 0-226-28088-8
39. Hilbert, David; and Ackermann, Wilhelm (1950); Principles of Mathematical Logic, Chelsea (English translation of Grundzüge der theoretischen Logik, 1928 German first edition)
40. Hodges, Wilfrid (2001); "Classical Logic I: First Order Logic", in Goble, Lou (ed.); The Blackwell Guide to Philosophical Logic, Blackwell
41. Ebbinghaus, Heinz-Dieter; Flum, Jörg; and Thomas, Wolfgang (1994); Mathematical Logic, Undergraduate Texts in Mathematics, Berlin, DE/New York, NY: Springer-Verlag, Second Edition, ISBN 978-0-387-94258-2
42. Rautenberg, Wolfgang (2010), A Concise Introduction to Mathematical Logic (3rd ed.), New York, NY: Springer Science+Business Media, doi:10.1007/978-1-4419-1221-3, ISBN 978-1-4419-1220-6
43. Tarski, Alfred and Givant, Steven (1987); A Formalization of Set Theory without Variables. Vol.41 of American Mathematical Society colloquium publications, Providence RI: American Mathematical Society, ISBN 978-0821810415.

CHAPTER 5

Degree of Truth

Chapter Contents

- **Half-Truth**
- **Quotations**
- **Truth Value**
- **Truth Value**
- **Vagueness**
- **Many-Valued Logic**
- **Three-Valued Logic**
- **Four-Valued Logic**
- **Truth Function**

This article does not cite any sources. Please help improve this article by adding citations to reliable sources. Unsourced material may be challenged and removed. (March 2009) (Learn how and when to remove this template message)

In standard mathematics, propositions can typically be considered unambiguously true or false. For instance, the proposition zero belongs to the set { 1 } is regarded as simply false; while the proposition one belongs to the set { 1 } is regarded as simply true. However, some mathematicians, computer scientists, and philosophers have been attracted to the idea that a proposition might be more or less true, rather than simply true or simply false. Consider My coffee is hot.

In mathematics, this idea can be developed in terms of fuzzy logic. In computer science, it has found application in artificial intelligence. In philosophy, the idea has proved particularly appealing in the case of vagueness. Degrees of truth is an important concept in law.

HALF-TRUTH

A **half-truth** is a deceptive statement that includes some element of truth. The statement might be partly true, the statement may be totally true but only part of the whole truth, or it may use some deceptive element, such as improper punctuation, or double meaning, especially if the intent is to deceive, evade, blame or misrepresent the truth.

Purpose

The purpose and or consequence of a half-truth is to make something that is really only a belief appear to be knowledge, or a truthful statement to represent the whole truth, or possibly lead to a false conclusion. According to the justified true belief theory of knowledge, in order to know that a given proposition is true, one must not only believe in the relevant true proposition, but one must also have a good reason for doing so. A half-truth deceives the recipient by presenting something believable and using those aspects of the statement that can be shown to be true as good reason to believe the statement is true in its entirety, or that the statement represents the whole truth. A person deceived by a half-truth considers the proposition to be knowledge and acts accordingly.

Examples

- "You should not trust Peter with your children. I once saw him smack a child with his open hand." In this example the statement could be true, but Peter may have slapped the child on the back because he was choking.
- "I'm a really good driver. In the past thirty years, I have gotten only four speeding tickets"* This statement is true, but irrelevant if the speaker started driving a week ago.
- Using a technicality: Bill Clinton famously engaged in a half-truth when he gave the testimony of "I did not have sexual relations with that woman, Miss Lewinsky." Here he engaged in an equivocation fallacy to deliberately indicate one particular meaning of the phrase "sexual relations", while intending another meaning, in order to deliberately mislead the court while still being able to later claim that "my statements were technically correct."
- After being stopped for drunk driving, the inebriated driver proclaims "I only had a couple of beers" in slurred speech. The driver may have also consumed alcoholic drinks other than beer, and the "beers" may have been large bottles as opposed to the usual contents of a normal-sized can, bottle, or glass
- The classic story about blind men and an elephant. Each blind man touches a different part of the elephant and reaches a different conclusion about the nature of the elephant; while each man's experience of the elephant is accurate, none of them have a full understanding of the nature of the beast. One may be touching the tail and believe that the elephant is long and thin, another may be touching the belly and say that it is round and big.
- "False dichotomy": the formal fallacy of false dilemma, also known as false choice, falsified dilemma, fallacy of the excluded middle, black and white

thinking, false correlative, either/or fallacy, and bifurcation—involves a situation in which two alternative statements are held to be the only possible options, when in reality there exist one or more other options which have not been considered or presented to the listeners.

Politics

Some forms of half-truths are an inescapable part of politics in representative democracies. The reputation of a political candidate can be irreparably damaged if they are exposed in a lie, so a complex style of language has evolved to minimise the chance of this happening. If someone has not said something, they cannot be accused of lying. As a consequence, politics has become a world where half-truths are expected, and political statements are rarely accepted at face value.

William Safire defines a half-truth, for political purposes, as "a statement accurate enough to require an explanation; and the longer the explanation, the more likely a public reaction of half-belief".

In his 1990 work The Magic Lantern: The Revolution of 1989 Witnessed in Warsaw, Budapest, Berlin, and Prague, Timothy Garton Ash responded to Václav Havel"s call for "living in truth": "Now we expect many things of politicians in a well-functioning parliamentary democracy. But "living in truth" is not one of them. In fact the essence of democratic politics might rather be described as "working in half-truth". Parliamentary democracy is, at its heart, a system of limited adversarial mendacity, in which each party attempts to present part of the truth as if it were the whole." Philosopher Alfred North Whitehead was quoted as saying: "There are no whole truths; all truths are half-truths. It is trying to treat them as whole truths that plays the devil". If this is true, statements, or truths, which according to Whitehead are all half-truths, are susceptible to creating deceptive and false conclusions.

Meme Theory

Richard Brodie links half-truths to memes "the truth of any proposition depends on the assumptions you make in considering it—the distinct memes you use in thinking about it". Brodie considers half-truths a necessary part of human interaction because they allow practical application of ideas when it is impractical to convey all the information needed to make a fully informed decision, although some half-truths can lead to a false conclusions or inferences in the world of logic.

QUOTATIONS

The notion of half-truths has existed in various cultures, giving rise to several epigrammatic sayings.

- Karl Kraus, an Austrian journalist, critic, playwright, and poet noted, "An aphorism can never be the whole truth; it is either a half-truth or a truth-and-a-half."

- Arthur Koestler "Two half-truths do not make a truth, and two half-cultures do not make a culture."

TRUTH

Truth is most often used to mean being in accord with fact or reality, or fidelity to an original or standard. Truth may also often be used in modern contexts to refer to an idea of "truth to self," or authenticity.

Time Saving Truth from Falsehood and Envy, François Lemoyne, 1737

Truth, holding a mirror and a serpent (1896). Olin Levi Warner, Library of Congress Thomas Jefferson Building, Washington, D.C.

The commonly understood opposite of truth is falsehood, which, correspondingly, can also take on a logical, factual, or ethical meaning. The concept of truth is discussed and debated in several contexts, including philosophy, art, and religion. Many human activities depend upon the concept, where its nature as a concept is assumed rather than being a subject of discussion; these include most (but not all) of the sciences, law, journalism, and everyday life. Some philosophers view the concept of truth as basic, and unable to be explained in any terms that are more easily understood than the concept of truth itself. Commonly, truth is viewed as the correspondence of language or thought to an independent reality, in what is sometimes called the correspondence theory of truth.

Other philosophers take this common meaning to be secondary and derivative. According to Martin Heidegger, the original meaning and essence of "Truth" in Ancient Greece was unconcealment, or the revealing or bringing of what was previously hidden into the open, as indicated by the original Greek term for truth, "Aletheia."On this view, the conception of truth as correctness is a later derivation from the concept"s original essence, a development Heidegger traces to the Latin term "Veritas."

Pragmatists like C.S. Peirce take Truth to have some manner of essential relation to human practices for inquiring into and discovering Truth, with Peirce himself holding that Truth is what human inquiry would find out on a matter, if our practice of inquiry were taken as far as it could profitably go: "The opinion which is fated to be ultimately agreed to by all who investigate, is what we mean by the truth..."

Various theories and views of truth continue to be debated among scholars, philosophers, and theologians. Language and words are a means by which humans convey information to one another and the method used to determine what is a "truth" is termed a criterion of truth. There are differing claims on such questions as what constitutes truth: what things are truthbearers capable of being true or false; how to define and identify truth; the roles that faith-based and empirically based knowledge play; and whether truth is subjective or objective, relative or absolute.

Friedrich Nietzsche famously suggested that an ancient, metaphysical belief in the divinity of Truth lies at the heart of and has served as the foundation for the entire subsequent Western intellectual tradition: "But you will have gathered what I am getting at, namely, that it is still a metaphysical faith on which our faith in science rests--that even we knowers of today, we godless anti-metaphysicians still take our fire too, from the flame lit by the thousand-year old faith, the Christian faith which was also Plato"s faith, that God is Truth; that Truth is ‹Divine'..."

Definition and Etymology

The English word truth is derived from Old English tríewþ, tréowþ, trýwþ, Middle English trewþe, cognate to Old High German triuwida, Old Norse tryggð. Like troth, it is a -th nominalisation of the adjective true (Old English tréowe).

An angel carrying the banner of "Truth", Roslin, Midlothian

The English word true is from Old English (West Saxon) (ge)tríewe, tréowe, cognate to Old Saxon (gi)trûui, Old High German (ga)triuwu (Modern German treu "faithful"), Old Norse tryggr, Gothic triggws, all from a Proto-Germanic *trewwj- "having good faith", perhaps ultimately from PIE *dru- "tree", on the notion of "steadfast as an oak" (e.g., Sanskrit "taru" tree). Old Norse trú, "faith, word of honour; religious faith, belief" (archaic English troth "loyalty, honesty, good faith", compare Ásatrú).

Thus, ‹truth" involves both the quality of "faithfulness, fidelity, loyalty, sincerity, veracity", and that of "agreement with fact or reality", in Anglo-Saxon expressed by sōþ (Modern English sooth).

All Germanic languages besides English have introduced a terminological distinction between truth "fidelity" and truth "factuality". To express "factuality", North Germanic opted for nouns derived from sanna "to assert, affirm", while continental West Germanic (German and Dutch) opted for continuations of wâra "faith, trust, pact" (cognate to Slavic věra "(religious) faith", but influenced by Latin verus). Romance languages use terms following the Latin veritas, while the Greek aletheia, Russian pravda and South Slavic istina have separate etymological origins.

Major Theories

The question of what is a proper basis for deciding how words, symbols, ideas and beliefs may properly be considered true, whether by a single person or an entire society, is dealt with by the five most prevalent substantive theories listed below. Each presents perspectives that are widely shared by published scholars.

However, the substantive theories are not universally accepted. More recently developed "deflationary" or "minimalist" theories of truth have emerged as competitors to the older substantive theories. Minimalist reasoning centres around the notion that the application of a term like true to a statement does not assert anything significant about it, for instance, anything about its nature. Minimalist reasoning realises truth as a label utilised in general discourse to express agreement, to stress claims, or to form general assumptions.

Substantive Theories

Correspondence Theory

Correspondence theories emphasise that true beliefs and true statements correspond to the actual state of affairs. This type of theory stresses a relationship between thoughts or statements on one hand, and things or objects on the other. It is a traditional model tracing its origins to ancient Greek philosophers such as Socrates, Plato, and Aristotle. This class of theories holds that the truth or the falsity of a representation is determined in principle entirely by how it relates to "things", by whether it accurately describes those "things." An example of correspondence theory is the statement by the thirteenth century philosopher/theologian Thomas Aquinas: Veritas est adaequatio rei et intellectus ("Truth is the equation [or adequation] of things and intellect"), a statement which Aquinas attributed to the ninth century neoplatonist Isaac Israeli. Aquinas also restated the theory as: "A judgment is said to be true when it conforms to the external reality".

Correspondence theory centres heavily around the assumption that truth is a matter of accurately copying what is known as "objective reality" and then representing it in thoughts, words and other symbols. Many modern theorists have stated that this ideal cannot be achieved without analysing additional factors. For example, language plays a role in that all languages have words to represent concepts that are virtually undefined in other languages. The German word Zeitgeist is one such example: one who speaks or understands the language may "know" what it means, but any translation of the word apparently fails to accurately capture its full meaning (this is a problem with many abstract words, especially those derived in agglutinative languages). Thus, some words add an additional parameter to the construction of an accurate truth predicate. Among the philosophers who grappled with this problem is Alfred Tarski, whose semantic theory is summarized further below in this article.

Proponents of several of the theories below have gone further to assert that there are yet other issues necessary to the analysis, such as interpersonal power struggles, community interactions, personal biases and other factors involved in deciding what is seen as truth.

Coherence Theory

Walter Seymour Allward's Veritas (Truth) outside Supreme Court of Canada, Ottawa, Ontario Canada

For coherence theories in general, truth requires a proper fit of elements within a whole system. Very often, though, coherence is taken to imply something more than simple logical consistency; often there is a demand that the propositions in a coherent system lend mutual inferential support to each other. So, for example, the completeness and comprehensiveness of the underlying set of concepts is a critical factor in judging the validity and usefulness of a coherent system. A pervasive tenet of coherence theories is the idea that truth is primarily a property of whole systems of propositions, and can be ascribed to individual propositions only according to their coherence with the whole. Among the assortment of perspectives commonly regarded as coherence theory, theorists differ on the question of whether coherence entails many possible true systems of thought or only a single absolute system.

Some variants of coherence theory are claimed to describe the essential and intrinsic properties of formal systems in logic and mathematics. However, formal reasoners are content to contemplate axiomatically independent and sometimes mutually contradictory systems side by side, for example, the various alternative geometries. On the whole, coherence theories have been rejected for lacking justification in their application to other areas of truth, especially with respect to assertions about the natural world, empirical data in general, assertions about practical matters of psychology and society, especially when used without support from the other major theories of truth.

Coherence theories distinguish the thought of rationalist philosophers, particularly of Spinoza, Leibniz, and G.W.F. Hegel, along with the British philosopher F.H. Bradley. They have found a resurgence also among several proponents of logical positivism, notably Otto Neurath and Carl Hempel.

Constructivist Theory

Social constructivism holds that truth is constructed by social processes, is historically and culturally specific, and that it is in part shaped through the power struggles within a community. Constructivism views all of our knowledge as "constructed," because it does not reflect any external "transcendent" realities (as a pure correspondence theory might hold). Rather, perceptions of truth are viewed as contingent on convention, human perception, and social experience. It is believed by constructivists that representations of physical and biological reality, including race, sexuality, and gender, are socially constructed.

Giambattista Vico was among the first to claim that history and culture were man-made. Vico's epistemological orientation gathers the most diverse rays and unfolds in one axiom – verum ipsum factum – "truth itself is constructed". Hegel and Marx were among the other early proponents of the premise that truth is, or can be, socially constructed. Marx, like many critical theorists who followed, did not reject the existence of objective truth but rather distinguished between true knowledge and knowledge that has been distorted through power or ideology. For Marx, scientific and true knowledge is "in accordance with the dialectical understanding of history" and ideological knowledge is "an epiphenomenal expression of the relation of material forces in a given economic arrangement".

Consensus Theory

Consensus theory holds that truth is whatever is agreed upon, or in some versions, might come to be agreed upon, by some specified group. Such a group might include all human beings, or a subset thereof consisting of more than one person.

Among the current advocates of consensus theory as a useful accounting of the concept of "truth" is the philosopher Jürgen Habermas. Habermas maintains that truth is what would be agreed upon in an ideal speech situation. Among the current strong critics of consensus theory is the philosopher Nicholas Rescher.

Pragmatic Theory

The three most influential forms of the pragmatic theory of truth were introduced around the turn of the 20th century by Charles Sanders Peirce, William James, and John Dewey. Although there are wide differences in viewpoint among these and other proponents of pragmatic theory, they hold in common that truth is verified and confirmed by the results of putting one"s concepts into practice.

Peirce defines truth as follows: "Truth is that concordance of an abstract statement with the ideal limit towards which endless investigation would tend to bring scientific belief, which concordance the abstract statement may possess by virtue of the confession of its inaccuracy and one-sidedness, and this confession is an essential ingredient of truth." This statement stresses Peirce's view that ideas of approximation, incompleteness, and partiality, what he describes elsewhere as fallibilism and "reference to the future", are essential to a proper conception of truth. Although Peirce uses words like concordance and correspondence to describe one aspect of the pragmatic sign relation, he is also quite explicit in saying that definitions of truth based on mere correspondence are no more than nominal definitions, which he accords a lower status than real definitions.

William James"s version of pragmatic theory, while complex, is often summarized by his statement that "the ‹true" is only the expedient in our way of thinking, just as the ‹right" is only the expedient in our way of behaving." By this, James meant that truth is a quality, the value of which is confirmed by its effectiveness when applying concepts to practice (thus, "pragmatic"). John Dewey, less broadly than James but more broadly than Peirce, held that inquiry, whether scientific, technical, sociological, philosophical or cultural, is self-corrective over time if openly submitted for testing by a community of inquirers in order to clarify, justify, refine and/or refute proposed truths. Though not widely known, a new variation of the pragmatic theory was defined and wielded successfully from the 20th century forward. Defined and named by William Ernest Hocking, this variation is known as "negative pragmatism". Essentially, what works may or may not be true, but what fails cannot be true because the truth always works. Richard Feynman also ascribed to it: "We never are definitely right, we can only be sure we are wrong." This approach incorporates many of the ideas from Peirce, James, and Dewey. For Peirce, the idea of "... endless investigation would tend to bring about scientific belief ..." fits negative pragmatism in that a negative pragmatist would never stop testing. As Feynman noted, an idea or theory "... could never be proved right, because tomorrow's experiment might succeed in proving wrong what you thought was right." Similarly, James and Dewey's ideas also ascribe truth to repeated testing which is "self-corrective" over time. Pragmatism and negative pragmatism are also closely aligned with the coherence theory of truth in that any testing should not be isolated but rather incorporate knowledge from all human endeavors and experience. The universe is a whole and integrated system, and testing should acknowledge and account for its diversity. As Feynman said, "... if it disagrees with experiment, it is wrong."

Minimalist (deflationary) Theories

Modern developments in the field of philosophy, starting with the relatively modern notion that a theory being old does not necessarily imply that it is completely flawless, have resulted in the rise of a new thesis: that the term truth does not denote a real property of sentences or propositions. This thesis is in part a response to the common use of truth predicates (e.g., that some particular thing "...is true") which was particularly prevalent in philosophical discourse on truth in the first half of the 20th century. From this point of view, to assert that "'2 + 2 = 4' is true" is logically equivalent to asserting that "2 + 2 = 4", and the phrase "is true" is completely dispensable in this and every other context. In common parlance, truth predicates are not commonly heard, and it would be interpreted as an unusual occurrence were someone to utilise a truth predicate in an everyday conversation when asserting that something is true. Newer perspectives that take this discrepancy into account and work with sentence structures that are actually employed in common discourse can be broadly described:

- as deflationary theories of truth, since they attempt to deflate the presumed importance of the words "true" or truth,
- as disquotational theories, to draw attention to the disappearance of the quotation marks in cases like the above example, or
- as minimalist theories of truth.

Whichever term is used, deflationary theories can be said to hold in common that "[t]he predicate ‹true" is an expressive convenience, not the name of a property requiring deep analysis." Once we have identified the truth predicate"s formal features and utility, deflationists argue, we have said all there is to be said about truth. Among the theoretical concerns of these views is to explain away those special cases where it does appear that the concept of truth has peculiar and interesting properties. (See, e.g., Semantic paradoxes, and below.)

In addition to highlighting such formal aspects of the predicate "is true", some deflationists point out that the concept enables us to express things that might otherwise require infinitely long sentences. For example, one cannot express confidence in Michael"s accuracy by asserting the endless sentence:

Michael says, 'snow is white' and snow is white, or he says 'roses are red' and roses are red or he says ... etc.

This assertion can also be succinctly expressed by saying: What Michael says is true.

Performative Theory of Truth

Attributed to P. F. Strawson is the performative theory of truth which holds that to say "'Snow is white' is true" is to perform the speech act of signaling one's agreement with the claim that snow is white (much like nodding one's head in agreement). The idea that some statements are more actions than communicative statements is not as odd as it may seem. Consider, for example, that when the bride says "I do" at the appropriate time in a wedding, she is performing the act of taking this man to be her lawful wedded husband. She is not describing herself as taking this man, but actually doing so (perhaps the most thorough analysis of such "illocutionary acts" is J. L. Austin, "How to Do Things With Words").

Strawson holds that a similar analysis is applicable to all speech acts, not just illocutionary ones: "To say a statement is true is not to make a statement about a statement, but rather to perform the act of agreeing with, accepting, or endorsing a statement. When one says ‹It"s true that it"s raining," one asserts no more than ‹It"s raining." The function of [the statement] ‹It"s true that..." is to agree with, accept, or endorse the statement that ‹it"s raining.""

Redundancy and Related Theories

According to the redundancy theory of truth, asserting that a statement is true is completely equivalent to asserting the statement itself. For example, making the assertion that " 'Snow is white' is true" is equivalent to asserting "Snow is white". Redundancy theorists infer from this premise that truth is a redundant concept; that is, it is merely a word that is traditionally used in conversation or writing, generally for emphasis, but not a word that actually equates to anything in reality. This theory is commonly attributed to Frank P. Ramsey, who held that the use of words like fact and truth was nothing but a roundabout way of asserting a proposition, and that treating these words as separate problems in isolation from judgment was merely a "linguistic muddle".

A variant of redundancy theory is the disquotational theory which uses a modified form of Tarski"s schema: To say that ‹"P" is true" is to say that P. A version of this theory was defended by C. J. F. Williams in his book What is Truth?. Yet another version of deflationism is the prosentential theory of truth, first developed by Dorothy Grover, Joseph Camp, and Nuel Belnap as an elaboration of Ramsey's claims. They argue that sentences like "That's true", when said in response to "It's raining", are prosentences, expressions that merely repeat the content of other expressions. In the same way that it means the same as my dog in the sentence My dog was hungry, so I fed it, That's true is supposed to mean the same as It"s raining — if you say the latter and I then say the former. These variations do not necessarily follow Ramsey in asserting that truth is not a property, but rather can be understood to say that, for instance, the assertion "P" may well involve a substantial truth, and the theorists in this case are minimizing only the redundancy or prosentence involved in the statement such as "that's true."

Deflationary principles do not apply to representations that are not analogous to sentences, and also do not apply to many other things that are commonly judged to be true or otherwise. Consider the analogy between the sentence "Snow is white" and the character named Snow White, both of which can be true in some sense. To a minimalist, saying "Snow is white is true" is the same as saying "Snow is white," but to say "Snow White is true" is not the same as saying "Snow White."

Pluralist Theories

Several of the major theories of truth hold that there is a particular property the having of which makes a belief or proposition true. Pluralist theories of truth assert that there may be more than one property that makes propositions true: ethical propositions might be true by virtue of coherence. Propositions about the physical world might be true by corresponding to the objects and properties they are about.

Some of the pragmatic theories, such as those by Charles Peirce and William James, included aspects of correspondence, coherence and constructivist theories. Crispin Wright argued in his 1992 book Truth and Objectivity that any predicate which satisfied certain platitudes about truth qualified as a truth predicate. In some discourses, Wright argued, the role of the truth predicate might be played by the notion of superassertibility. Michael Lynch, in a 2009 book Truth as One and Many, argued that we should see truth as a functional property capable of being multiply manifested in distinct properties like correspondence or coherence.

Most Believed Theories

According to a survey of professional philosophers and others on their philosophical views which was carried out in November 2009 (taken by 3226 respondents, including 1803 philosophy faculty members and/or PhDs and 829 philosophy graduate students) 44.9% of respondents accept or lean towards correspondence theories, 20.7% accept or lean towards deflationary theories and 13.8% epistemic theories.

Formal Theories

Truth in Logic

Logic is concerned with the patterns in reason that can help tell us if a proposition is true or not. However, logic does not deal with truth in the absolute sense, as for instance a metaphysician does. Logicians use formal languages to express the truths which they are concerned with, and as such there is only truth under some interpretation or truth within some logical system.

A logical truth (also called an analytic truth or a necessary truth) is a statement which is true in all possible worlds or under all possible interpretations, as contrasted to a fact (also called a synthetic claim or a contingency) which is only true in this world as it has historically unfolded. A proposition such as "If p and q, then p" is considered to be a logical truth because of the meaning of the symbols and words in it and not because of any fact of any particular world. They are such that they could not be untrue.

Truth in Mathematics

There are two main approaches to truth in mathematics. They are the model theory of truth and the proof theory of truth.

Historically, with the nineteenth century development of Boolean algebra mathematical models of logic began to treat "truth", also represented as "T" or "1", as an arbitrary constant. "Falsity" is also an arbitrary constant, which can be represented as "F" or "0". In propositional logic, these symbols can be manipulated according to a set of axioms and rules of inference, often given in the form of truth tables.

In addition, from at least the time of Hilbert"s program at the turn of the twentieth century to the proof of Gödel"s incompleteness theorems and the development of the Church-Turing thesis in the early part of that century, true statements in mathematics were generally assumed to be those statements that are provable in a formal axiomatic system.

The works of Kurt Gödel, Alan Turing, and others shook this assumption, with the development of statements that are true but cannot be proven within the system. Two examples of the latter can be found in Hilbert"s problems. Work on Hilbert"s 10th problem led in the late twentieth century to the construction of specific Diophantine equations for which it is undecidable whether they have a solution, or even if they do, whether they have a finite or infinite number of solutions. More fundamentally, Hilbert"s first problem was on the continuum hypothesis. Gödel and Paul Cohen showed that this hypothesis cannot be proved or disproved using the standard axioms of set theory. In the view of some, then, it is equally reasonable to take either the continuum hypothesis or its negation as a new axiom.

Semantic Theory of Truth

The semantic theory of truth has as its general case for a given language:

"P" is true if and only if P

where ‹P" refers to the sentence (the sentence"s name), and P is just the sentence itself.

Logician and philosopher Alfred Tarski developed the theory for formal languages (such as formal logic). Here he restricted it in this way: no language could contain its own truth predicate, that is, the expression is true could only apply to sentences in some other language. The latter he called an object language, the language being talked about. (It may, in turn, have a truth predicate that can be applied to sentences in still another language.) The reason for his restriction was that languages that contain their own truth predicate will contain paradoxical sentences such as, "This sentence is not true". As a result, Tarski held that the semantic theory could not be applied to any natural language, such as English, because they contain their own truth predicates. Donald Davidson used it as the foundation of his truth-conditional semantics and linked it to radical interpretation in a form of coherentism.

Bertrand Russell is credited with noticing the existence of such paradoxes even in the best symbolic formations of mathematics in his day, in particular the paradox that came to be named after him, Russell"s paradox. Russell and Whitehead attempted to solve these problems in Principia Mathematica by putting statements into a hierarchy of types, wherein a statement cannot refer to itself, but only to statements lower in the hierarchy. This in turn led to new orders of difficulty regarding the precise natures of types and the structures of conceptually possible type systems that have yet to be resolved to this day.

Kripke"s Theory of Truth

Saul Kripke contends that a natural language can in fact contain its own truth predicate without giving rise to contradiction. He showed how to construct one as follows:

- Begin with a subset of sentences of a natural language that contains no occurrences of the expression "is true" (or "is false"). So The barn is big is included in the subset, but not " The barn is big is true", nor problematic sentences such as "This sentence is false".
- Define truth just for the sentences in that subset.
- Then extend the definition of truth to include sentences that predicate truth or falsity of one of the original subset of sentences. So "The barn is big is true" is now included, but not either "This sentence is false" nor "'The barn is big is true' is true".
- Next, define truth for all sentences that predicate truth or falsity of a member of the second set. Imagine this process repeated infinitely, so that truth is defined for The barn is big; then for "The barn is big is true"; then for "'The barn is big is true' is true", and so on.

Notice that truth never gets defined for sentences like This sentence is false, since it was not in the original subset and does not predicate truth of any sentence in the original or any subsequent set. In Kripke's terms, these are "ungrounded." Since these sentences are never assigned either truth or falsehood even if the process is carried out infinitely, Kripke's theory implies that some sentences are neither true nor false. This contradicts the Principle of bivalence: every sentence must be either true or false. Since this principle is a key premise in deriving the Liar paradox, the paradox is dissolved.

However, it has been shown by Gödel that self-reference cannot be avoided naively, since propositions about seemingly unrelated objects can have an informal self-referential meaning; in Gödel's work, these objects are integer numbers, and they have an informal meaning regarding propositions. In fact, this idea - manifested by the diagonal lemma - is the basis for Tarski"s theorem that truth cannot be consistently defined.

It has thus been claimed that Kripke"s system indeed leads to contradiction: while its truth predicate is only partial, it does give truth value (true/false) to propositions such as the one built in Tarski"s proof, and is therefore inconsistent. While there is still a debate on whether Tarski"s proof can be implemented to every similar partial truth system, none have been shown to be consistent by acceptable methods used in mathematical logic.

Revision Theory of Truth

The revision theory of truth, as developed by Anil Gupta and Nuel Belnap, takes truth to be a circular concept whose definition is the set of biconditionals of the form

'A' is true if and only if A.

Unlike Kripke's theory of truth, revision theory can be used with classical logic and can maintain the principle of bivalence.

Notable views

La Vérité "Truth" by Jules Joseph Lefebvre

Ancient History

The ancient Greek origins of the words "true" and "truth" have some consistent definitions throughout great spans of history that were often associated with topics of logic, geometry, mathematics, deduction, induction, and natural philosophy.

Socrates", Plato"s and Aristotle"s ideas about truth are seen by some as consistent with correspondence theory. In his Metaphysics, Aristotle stated: "To say of what is that it is not, or of what is not that it is, is false, while to say of what is that it is, and of what is not that it is not, is true". The Stanford Encyclopedia of Philosophy proceeds to say of Aristotle:

"(...) Aristotle sounds much more like a genuine correspondence theorist in the Categories (12b11, 14b14), where he talks of "underlying things" that make statements true and implies that these "things" (pragmata) are logically structured situations or facts (viz., his sitting, his not sitting). Most influential is his claim in De Interpretatione (16a3) that thoughts are "likenesses" (homoiosis) of things. Although he nowhere defines truth in terms of a thought"s likeness to a thing or fact, it is clear that such a definition would fit well into his overall philosophy of mind. (...)"

Very similar statements can also be found in Plato (Cratylus 385b2, Sophist 263b).

In Hinduism, Truth is defined as "unchangeable", "that which has no distortion", "that which is beyond distinctions of time, space, and person", "that which pervades the universe in all its constancy". The human body, therefore is not completely true as it changes with time, for example. There are many references, properties and explanations of truth by Hindu sages that explain varied facets of truth, such as the national motto of India: "Satyameva jayate" (Truth alone wins), as well as "Satyam muktaye" (Truth liberates), "Satya' is 'Parahit'artham' va'unmanaso yatha'rthatvam' satyam" (Satya is the benevolent use of words and the mind for the welfare of others or in other words responsibilities is truth too), "When one is firmly established in speaking truth, the fruits of action become subservient to him (patanjali yogasutras, sutra number 2.36), "The face of truth is covered by a golden bowl. Unveil it, O Pusan (Sun), so that I who have truth as my duty (satyadharma) may see it!" (Brhadaranyaka V 15 1-4 and the brief IIsa Upanisad 15-18), Truth is superior to silence (Manusmriti), etc. Combined with other words, satya acts as modifier, like "ultra" or "highest," or more literally "truest," connoting purity and excellence. For example, satyaloka is the "highest heaven' and Satya Yuga is the "golden age" or best of the four cyclical cosmic ages in Hinduism, and so on.

Middle Ages

Avicenna

In early Islamic philosophy, Avicenna (Ibn Sina) defined truth in his work Kitab Al-Shifa The Book of Healing, Book I, Chapter 8, as:

"What corresponds in the mind to what is outside it."

Avicenna elaborated on his definition of truth later in Book VIII, Chapter 6:

"The truth of a thing is the property of the being of each thing which has been established in it."

However, this definition is merely a rendering of the medieval Latin translation of the work by Simone van Riet. A modern translation of the original Arabic text states:

"Truth is also said of the veridical belief in the existence [of something]".

Aquinas

Reevaluating Avicenna, and also Augustine and Aristotle, Thomas Aquinas stated in his Disputed Questions on Truth:

A natural thing, being placed between two intellects, is called true insofar as it conforms to either. It is said to be true with respect to its conformity with the divine intellect insofar as it fulfills the end to which it was ordained by the divine intellect... With respect to its conformity with a human intellect, a thing is said to be true insofar as it is such as to cause a true estimate about itself.

Thus, for Aquinas, the truth of the human intellect (logical truth) is based on the truth in things (ontological truth). Following this, he wrote an elegant re-statement of Aristotle"s view in his Summa I.16.1:

Veritas est adæquatio intellectus et rei.

(Truth is the conformity of the intellect to the things.)

Aquinas also said that real things participate in the act of being of the Creator God who is Subsistent Being, Intelligence, and Truth. Thus, these beings possess the light of intelligibility and are knowable. These things (beings; reality) are the foundation of the truth that is found in the human mind, when it acquires knowledge of things, first through the senses, then through the understanding and the judgement done by reason. For Aquinas, human intelligence ("intus", within and "legere", to read) has the capability to reach the essence and existence of things because it has a non-material, spiritual element, although some moral, educational, and other elements might interfere with its capability.

Changing concepts of truth in the Middle Ages

Richard Firth Green examined the concept of truth in the later Middle Ages in his A Crisis of Truth, and concludes that roughly during the reign of Richard II of England the very meaning of the concept changes. The idea of the oath, which was so much part and parcel of for instance Romance literature, changes from a subjective concept to a more objective one (in Derek Pearsall"s summary). Whereas truth (the "trouthe" of Sir Gawain and the Green Knight) was first "an ethical truth in which truth is understood to reside in persons", in Ricardian England it "transforms...into a political truth in which truth is understood to reside in documents".

Modern age

Kant

Immanuel Kant

Immanuel Kant endorses a definition of truth along the lines of the correspondence theory of truth. Kant writes in the Critique of Pure Reason: "The nominal definition of truth, namely that it is the agreement of cognition with its object, is here granted and presupposed". However, Kant denies that this correspondence definition of truth provides us with a test or criterion to establish which judgements are true. Kant states in his logic lectures:

"(...) Truth, it is said, consists in the agreement of cognition with its object. In consequence of this mere nominal definition, my cognition, to count as true, is supposed to agree with its object. Now I can compare the object with my cognition, however, only by cognizing it. Hence my cognition is supposed to confirm itself, which is far short of being sufficient for truth. For since the object is outside me, the cognition in me, all I can ever pass judgement on is whether my cognition of the object agrees with my cognition of the object. The ancients called such a circle in explanation a diallelon. And actually the logicians were always reproached with this mistake by the sceptics, who observed that with this definition of truth it is just as when someone makes a statement before a court and in doing so appeals to a witness with whom no one is acquainted, but who wants to establish his credibility by maintaining that the one who called him as witness is an honest man. The accusation was grounded, too. Only the solution of the indicated problem is impossible without qualification and for every man. (...)"

This passage makes use of his distinction between nominal and real definitions. A nominal definition explains the meaning of a linguistic expression. A real definition describes the essence of certain objects and enable us to determine whether any given item falls within the definition. Kant holds that the definition of truth is merely nominal and, therefore, we cannot employ it to establish which judgements are true. According to Kant, the ancient skeptics were critical of the logicians for holding that, by means of a merely nominal definition of truth, they can establish which judgements are true. They were trying to do something that is "impossible without qualification and for every man".

Hegel

Georg Hegel distanced his philosophy from psychology by presenting truth as being an external self-moving object instead of being related to inner, subjective thoughts. Hegel's truth is analogous to the mechanics of a material body in motion under the influence of its own inner force. "Truth is its own self-movement within itself." Teleological truth moves itself in the three-step form of dialectical triplicity toward the final goal of perfect, final, absolute truth. According to Hegel, the progression of philosophical truth is a resolution of past oppositions into increasingly more accurate approximations of absolute truth. Chalybäus used the terms "thesis", "antithesis", and "synthesis" to describe Hegel"s dialectical triplicity. The "thesis" consists of an incomplete historical movement. To resolve the incompletion, an "antithesis" occurs which opposes the "thesis." In turn, the "synthesis" appears when the "thesis" and "antithesis" become reconciled and a higher level of truth is obtained. This "synthesis" thereby becomes a "thesis," which will again necessitate an "antithesis," requiring a new "synthesis" until a final state is reached as the result of reason's historical movement. History is the Absolute Spirit moving toward a goal. This historical progression will finally conclude itself when the Absolute Spirit understands its own infinite self at the very end of history. Absolute Spirit will then be the complete expression of an infinite God.

Schopenhauer

For Arthur Schopenhauer, a judgment is a combination or separation of two or more concepts. If a judgment is to be an expression of knowledge, it must have a sufficient reason or ground by which the judgment could be called true. Truth is the reference of a judgment to something different from itself which is its sufficient reason (ground). Judgments can have material, formal, transcendental, or metalogical truth. A judgment has material truth if its concepts are based on intuitive perceptions that are generated from sensations. If a judgment has its reason (ground) in another judgment, its truth is called logical or formal. If a judgment, of, for example, pure mathematics or pure science, is based on the forms (space, time, causality) of intuitive, empirical knowledge, then the judgment has transcendental truth.

Kierkegaard

When Søren Kierkegaard, as his character Johannes Climacus, ends his writings: My thesis was, subjectivity, heartfelt is the truth, he does not advocate for subjectivism in its extreme form (the theory that something is true simply because one believes it to be so), but rather that the objective approach to matters of personal truth cannot shed any light upon that which is most essential to a person's life. Objective truths are concerned with the facts of a person's being, while subjective truths are concerned with a person's way of being. Kierkegaard agrees that objective truths for the study of subjects like mathematics, science, and history are relevant and necessary, but argues that objective truths do not shed any light on a person's inner relationship to existence. At best, these truths can only provide a severely narrowed perspective that has little to do with one's actual experience of life.

While objective truths are final and static, subjective truths are continuing and dynamic. The truth of one"s existence is a living, inward, and subjective experience that is always in the process of becoming. The values, morals, and spiritual approaches a person adopts, while not denying the existence of objective truths of those beliefs, can only become truly known when they have been inwardly appropriated through subjective experience. Thus, Kierkegaard criticizes all systematic philosophies which attempt to know life or the truth of existence via theories and objective knowledge about reality. As Kierkegaard claims, human truth is something that is continually occurring, and a human being cannot find truth separate from the subjective experience of one"s own existing, defined by the values and fundamental essence that consist of one"s way of life.

Nietzsche

Friedrich Nietzsche believed the search for truth, or 'the will to truth', was a consequence of the will to power of philosophers. He thought that truth should be used as long as it promoted life and the will to power, and he thought untruth was better than truth if it had this life enhancement as a consequence. As he wrote in Beyond Good and Evil, "The falseness of a judgment is to us not necessarily an objection to a judgment... The question is to what extent it is life-advancing, life-preserving, species-preserving, perhaps even species-breeding..." (aphorism 4). He proposed the will to power as a truth only because, according to him, it was the most life-affirming and sincere perspective one could have.

Robert Wicks discusses Nietzsche"s basic view of truth as follows:

"(...) Some scholars regard Nietzsche"s 1873 unpublished essay, "On Truth and Lies in a Nonmoral Sense" ("Über Wahrheit und Lüge im außermoralischen Sinn") as a keystone in his thought. In this essay, Nietzsche rejects the idea of universal constants, and claims that what we call "truth" is only "a mobile army of metaphors, metonyms, and anthropomorphisms." His view at this time is that arbitrariness completely prevails within human experience: concepts originate via the very artistic transference of nerve stimuli into images; "truth" is nothing more than the invention of fixed conventions for merely practical purposes, especially those of repose, security and consistence. (...)"

Whitehead

Alfred North Whitehead, a British mathematician who became an American philosopher, said: "There are no whole truths; all truths are half-truths. It is trying to treat them as whole truths that plays the devil".

The logical progression or connection of this line of thought is to conclude that truth can lie, since half-truths are deceptive and may lead to a false conclusion.

Nishida

According to Kitaro Nishida, "knowledge of things in the world begins with the differentiation of unitary consciousness into knower and known and ends with self and things becoming one again. Such unification takes form not only in knowing but in the valuing (of truth) that directs knowing, the willing that directs action, and the feeling or emotive reach that directs sensing."

Fromm

Erich Fromm finds that trying to discuss truth as "absolute truth" is sterile and that emphasis ought to be placed on "optimal truth". He considers truth as stemming from the survival imperative of grasping one's environment physically and intellectually, whereby young children instinctively seek truth so as to orient themselves in "a strange and powerful world". The accuracy of their perceived approximation of the truth will therefore have direct consequences on their ability to deal with their environment. Fromm can be understood to define truth as a functional approximation of reality. His vision of optimal truth is described partly in "Man from Himself: An Inquiry into the Psychology of Ethics" (1947), from which excerpts are included below. the dichotomy between ‹absolute = perfect" and ‹relative = imperfect" has been superseded in all fields of scientific thought, where "it is generally recognized that there is no absolute truth but nevertheless that there are objectively valid laws and principles".

In that respect, "a scientifically or rationally valid statement means that the power of reason is applied to all the available data of observation without any of them being suppressed or falsified for the sake of a desired result". The history of science is "a history of inadequate and incomplete statements, and every new insight makes possible the recognition of the inadequacies of previous propositions and offers a springboard for creating a more adequate formulation."

As a result "the history of thought is the history of an ever-increasing approximation to the truth. Scientific knowledge is not absolute but optimal; it contains the optimum of truth attainable in a given historical period." Fromm furthermore notes that "different cultures have emphasized various aspects of the truth" and that increasing interaction between cultures allows for these aspects to reconcile and integrate, increasing further the approximation to the truth.

Foucault

Quod Est Veritas? Christ and Pilate, by Nikolai Ge.

Truth, says Michel Foucault, is problematic when any attempt is made to see truth as an "objective" quality. He prefers not to use the term truth itself but "Regimes of Truth". In his historical investigations he found truth to be something that was itself a part of, or embedded within, a given power structure. Thus Foucault"s view shares much in common with the concepts of Nietzsche. Truth for Foucault is also something that shifts through various episteme throughout history.

Baudrillard

Jean Baudrillard considered truth to be largely simulated, that is pretending to have something, as opposed to dissimulation, pretending to not have something. He took his cue from iconoclasts who he claims knew that images of God demonstrated that God did not exist. Baudrillard wrote in "Precession of the Simulacra":

The simulacrum is never that which conceals the truth—it is the truth which conceals that there is none. The simulacrum is true.

—Ecclesiastes

Some examples of simulacra that Baudrillard cited were: that prisons simulate the "truth" that society is free; scandals (e.g., Watergate) simulate that corruption is corrected; Disney simulates that the U.S. itself is an adult place. One must remember that though such examples seem extreme, such extremity is an important part of Baudrillard"s theory. For a less extreme example, consider how movies usually end with the bad being punished, humiliated, or otherwise failing, thus affirming for viewers the concept that the good end happily and the bad unhappily, a narrative which implies that the status quo and established power structures are largely legitimate.

In Medicine and Psychiatry

There is controversy as to the truth value of a proposition made in bad faith self-deception, such as when a hypochondriac has a complaint with no physical symptom.

In religion: Omniscience

In a religious context, perfect knowledge of all truth about all things (omniscience) is regarded by some religions, particularly Buddhism (source?) and the Abrahamic religions (Christianity, Islam, and Judaism), as an attribute of a divine being. In the Abrahamic view, God can exercise divine judgment, judging the dead on the basis of perfect knowledge of their lives.

TRUTH VALUE

In logic and mathematics, a **truth value**, sometimes called a **logical value**, is a value indicating the relation of a proposition to truth.

Classical logic

In classical logic, with its intended semantics, the truth values are **true** (1 or T) and **untrue** or **false** (0 or ⊥); that is, classical logic is a two-valued logic. This set of two values is also called the Boolean domain. Corresponding semantics of logical connectives are truth functions, whose values are expressed in the form of truth tables. Logical biconditional becomes the equality binary relation, and negation becomes a bijection which permutes true and false. Conjunction and disjunction are dual with respect to negation, which is expressed by De Morgan"s laws:

$\neg(p\wedge q) \Leftrightarrow \neg p \vee \neg q$

$\neg(p\vee q) \Leftrightarrow \neg p \wedge \neg q$

Propositional variables become variables in the Boolean domain. Assigning values for propositional variables is referred to as valuation.

Intuitionistic and Constructive Logic

In intuitionistic logic, and more generally, constructive mathematics, statements are assigned a truth value only if they can be given a constructive proof. It starts with a set of axioms, and a statement is true if you can build a proof of the statement from those axioms. A statement is false if you can deduce a contradiction from it. This leaves open the possibility of statements that have not yet been assigned a truth value. Unproved statements in Intuitionistic logic are not given an intermediate truth value (as is sometimes mistakenly asserted). Indeed, you can prove that they have no third truth value, a result dating back to Glivenko in 1928

Instead statements simply remain of unknown truth value, until they are either proved or disproved.

There are various ways of interpreting Intuitionistic logic, including the Brouwer–Heyting–Kolmogorov interpretation. See also, Intuitionistic Logic - Semantics.

In logic and mathematics, a **truth value**, sometimes called a **logical value**, is a value indicating the relation of a proposition to truth.

In classical logic, with its intended semantics, the truth values are **true** (1 or T) and **untrue** or **false** (0 or ⊥); that is, classical logic is a two-valued logic. This set of two values is also called the Boolean domain. Corresponding semantics of logical connectives are truth functions, whose values are expressed in the form of truth tables. Logical biconditional becomes the equality binary relation, and negation becomes a bijection which permutes true and false. Conjunction and disjunction are dual with respect to negation, which is expressed by De Morgan"s laws:

$\neg(p \wedge q) \Leftrightarrow \neg p \vee \neg q$

$\neg(p \vee q) \Leftrightarrow \neg p \wedge \neg q$

Propositional variables become variables in the Boolean domain. Assigning values for propositional variables is referred to as valuation.

Intuitionistic and Constructive Logic

In intuitionistic logic, and more generally, constructive mathematics, statements are assigned a truth value only if they can be given a constructive proof. It starts with a set of axioms, and a statement is true if you can build a proof of the statement from those axioms. A statement is false if you can deduce a contradiction from it. This leaves open the possibility of statements that have not yet been assigned a truth value. Unproved statements in Intuitionistic logic are not given an intermediate truth value (as is sometimes mistakenly asserted). Indeed, you can prove that they have no third truth value, a result dating back to Glivenko in 1928

Instead statements simply remain of unknown truth value, until they are either proved or disproved.

There are various ways of interpreting Intuitionistic logic, including the Brouwer–Heyting–Kolmogorov interpretation. See also, Intuitionistic Logic - Semantics.

Multi-valued Logic

Multi-valued logics (such as fuzzy logic and relevance logic) allow for more than two truth values, possibly containing some internal structure. For example, on the unit interval [0,1] such structure is a total order; this may be expressed as existence of various degrees of truth.

Not all logical systems are truth-valuational in the sense that logical connectives may be interpreted as truth functions. For example, intuitionistic logic lacks a complete set of truth values because its semantics, the Brouwer–Heyting–Kolmogorov interpretation, is specified in terms of provability conditions, and not directly in terms of the necessary truth of formulae.

But even non-truth-valuational logics can associate values with logical formulae, as is done in algebraic semantics. The algebraic semantics of intuitionistic logic is given in terms of Heyting algebras, compared to Boolean algebra semantics of classical propositional calculus.

Intuitionistic type theory uses types in the place of truth values.

Topos theory uses truth values in a special sense: the truth values of a topos are the global elements of the subobject classifier. Having truth values in this sense does not make a logic truth valuational.

Multi-valued logics (such as fuzzy logic and relevance logic) allow for more than two truth values, possibly containing some internal structure. For example, on the unit interval [0,1] such structure is a total order; this may be expressed as existence of various degrees of truth.

Algebraic Semantics

Not all logical systems are truth-valuational in the sense that logical connectives may be interpreted as truth functions. For example, intuitionistic logic lacks a complete set of truth values because its semantics, the Brouwer–Heyting–Kolmogorov interpretation, is specified in terms of provability conditions, and not directly in terms of the necessary truth of formulae.

But even non-truth-valuational logics can associate values with logical formulae, as is done in algebraic semantics. The algebraic semantics of intuitionistic logic is given in terms of Heyting algebras, compared to Boolean algebra semantics of classical propositional calculus.

In other Theories

Intuitionistic type theory uses types in the place of truth values.

Topos theory uses truth values in a special sense: the truth values of a topos are the global elements of the subobject classifier. Having truth values in this sense does not make a logic truth valuational.

VAGUENESS

In analytic philosophy and linguistics, a concept may be considered **vague** if its extension is deemed lacking in clarity, if there is uncertainty about which objects belong to the concept or which exhibit characteristics that have this predicate (so-called "border-line cases"), or if the Sorites paradox applies to the concept or predicate.

In everyday speech, vagueness is an inevitable, often even desired effect of language usage. However, in most specialized texts (e.g., legal documents), vagueness is often regarded as problematic and undesirable.

Importance

Vagueness is philosophically important. Suppose one wants to come up with a definition of "right" in the moral sense. One wants a definition to cover actions that are clearly right and exclude actions that are clearly wrong, but what does one do with the borderline cases? Surely, there are such cases. Some philosophers say that one should try to come up with a definition that is itself unclear on just those cases. Others say that one has an interest in making his or her definitions more precise than ordinary language, or his or her ordinary concepts, themselves allow; they recommend one advances precising definitions.

Vagueness is also a problem which arises in law, and in some cases judges have to arbitrate regarding whether a borderline case does, or does not, satisfy a given vague concept. Examples include disability (how much loss of vision is required before one is legally blind?), human life (at what point from conception to birth is one a legal human being, protected for instance by laws against murder?), adulthood (most familiarly reflected in legal ages for driving, drinking, voting, consensual sex, etc.), race (how to classify someone of mixed racial heritage), etc. Even such apparently unambiguous concepts such as gender can be subject to vagueness problems, not just from transsexuals' gender transitions but also from certain genetic conditions which can give an individual mixed male and female biological traits (see intersex).

Many scientific concepts are of necessity vague, for instance species in biology cannot be precisely defined, owing to unclear cases such as ring species. Nonetheless, the concept of species can be clearly applied in the vast majority of cases. As this example illustrates, to say that a definition is "vague" is not necessarily a criticism. Consider those animals in Alaska that are the result of breeding huskies and wolves: are they dogs? It is not clear: they are borderline cases of dogs. This means one"s ordinary concept of doghood is not clear enough to let us rule conclusively in this case.

Approaches

The philosophical question of what the best theoretical treatment of vagueness is - which is closely related to the problem of the paradox of the heap, a.k.a. sorites paradox - has been the subject of much philosophical debate.

Fuzzy logic

One theoretical approach is that of fuzzy logic, developed by American mathematician Lotfi Zadeh. Fuzzy logic proposes a gradual transition between "perfect falsity", for example, the statement "Bill Clinton is bald", to "perfect truth", for, say, "Patrick Stewart is bald". In ordinary logics, there are only two truth-values: "true" and "false". The fuzzy perspective differs by introducing an infinite number of truth-values along a spectrum between perfect truth and perfect falsity. Perfect truth may be represented by "1", and perfect falsity by "0". Borderline cases are thought of as having a "truth-value" anywhere between 0 and 1 (for example, 0.6). Advocates of the fuzzy logic approach have included K. F. Machina (1976) and Dorothy Edgington (1993).

Supervaluationism

Another theoretical approach is known as "supervaluationism". This approach has been defended by Kit Fine and Rosanna Keefe. Fine argues that borderline applications of vague predicates are neither true nor false, but rather are instances of "truth value gaps". He defends an interesting and sophisticated system of vague semantics, based on the notion that a vague predicate might be "made precise" in many alternative ways. This system has the consequence that borderline cases of vague terms yield statements that are neither true, nor false.

Given a supervaluationist semantics, one can define the predicate ‹supertrue" as meaning "true on all precisifications". This predicate will not change the semantics of atomic statements (e.g. ‹Frank is bald", where Frank is a borderline case of baldness), but does have consequences for logically complex statements. In particular, the tautologies of sentential logic, such as 'Frank is bald or Frank is not bald', will turn out to be supertrue, since on any precisification of baldness, either 'Frank is bald' or 'Frank is not bald' will be true. Since the presence of borderline cases seems to threaten principles like this one (excluded middle), the fact that supervaluationism can "rescue" them is seen as a virtue.

The Epistemicist View

A third approach, known as the "epistemicist view", has been defended by Timothy Williamson (1994), R. A. Sorensen (1988) and (2001), and Nicholas Rescher (2009). They maintain that vague predicates do, in fact, draw sharp boundaries, but that one cannot know where these boundaries lie. One's confusion about whether some vague word does or does not apply in a borderline case is explained as being due to one"s ignorance. For example, on the epistemicist view, there is a fact of the matter, for every person, about whether that person is old or not old. It is just that one may sometimes be ignorant of this fact.

Vagueness as a Property of Objects

One possibility is that one's words and concepts are perfectly precise, but that objects themselves are vague. Consider Peter Unger's example of a cloud (from his famous 1980 paper, "The Problem of the Many") : it's not clear where the boundary of a cloud lies; for any given bit of water vapor, one can ask whether it's part of the cloud or not, and for many such bits, one won't know how to answer. So perhaps one's term 'cloud' denotes a vague object precisely. This strategy has been poorly received, in part due to Gareth Evans"s short paper "Can There Be Vague Objects?" (1978). Evans's argument appears to show that there can be no vague identities (e.g. "Princeton = Princeton Borough"), but as Lewis (1988) makes clear, Evans takes for granted that there are in fact vague identities, and that any proof to the contrary cannot be right. Since the proof Evans produces relies on the assumption that terms precisely denote vague objects, the implication is that the assumption is false, and so the vague-objects view is wrong.

Still by, for instance, proposing alternative deduction rules involving Leibniz"s law or other rules for validity some philosophers are willing to defend vagueness as some kind of metaphysical phenomenon. One has, for example, Peter van Inwagen (1990), Trenton Merricks and Terence Parsons (2000).

Legal Principle

In the common law system, vagueness is a possible legal defence against by-laws and other regulations. The legal principle is that delegated power cannot be used more broadly than the delegator intended. Therefore, a regulation may not be so vague as to regulate areas beyond what the law allows. Any such regulation would be "void for vagueness" and unenforceable. This principle is sometimes used to strike down municipal by-laws that forbid "explicit" or "objectionable" contents from being sold in a certain city; courts often find such expressions to be too vague, giving municipal inspectors discretion beyond what the law allows. In the US this is known as the vagueness doctrine and in Europe as the principle of legal certainty.

MANY-VALUED LOGIC

In logic, a many-valued logic (also multi- or multiple-valued logic) is a propositional calculus in which there are more than two truth values. Traditionally, in Aristotle"s logical calculus, there were only two possible values (i.e., "true" and "false") for any proposition. Classical two-valued logic may be extended to n-valued logic for n greater than 2. Those most popular in the literature are three-valued (e.g., Łukasiewicz's and Kleene"s, which accept the values "true", "false", and "unknown"), the finite-valued (finitely-many valued) with more than three values, and the infinite-valued (infinitely-many valued), such as fuzzy logic and probability logic. The first known classical logician who didn"t fully accept the law of excluded middle was Aristotle (who, ironically, is also generally considered to be the first classical logician and the "father of logic"). Aristotle admitted that his laws did not all apply to future events (De Interpretatione, ch. IX), but he didn't create a system of multi-valued logic to explain this isolated remark. Until the coming of the 20th century, later logicians followed Aristotelian logic, which includes or assumes the law of the excluded middle.

The 20th century brought back the idea of multi-valued logic. The Polish logician and philosopher Jan Łukasiewicz began to create systems of many-valued logic in 1920, using a third value, "possible", to deal with Aristotle's paradox of the sea battle. Meanwhile, the American mathematician, Emil L. Post (1921), also introduced the formulation of additional truth degrees with $n \geq 2$, where n are the truth values. Later, Jan Łukasiewicz and Alfred Tarski together formulated a logic on n truth values where $n \geq 2$. In 1932 Hans Reichenbach formulated a logic of many truth values where $n \rightarrow \infty$. Kurt Gödel in 1932 showed that intuitionistic logic is not a finitely-many valued logic, and defined a system of Gödel logics intermediate between classical and intuitionistic logic; such logics are known as intermediate logics.

Examples

Kleene (strong) K_3 *and Priest logic* P_3

Kleene's "(strong) logic of indeterminacy" K_3 (sometimes K_3^S) and Priest's "logic of paradox" add a third "undefined" or "indeterminate" truth value I. The truth functions for negation (¬), conjunction (∧), disjunction (∨), implication (→K), and biconditional (↔K) are given by:

¬	
T	F
I	I
F	T

∧	T	I	F
T	T	I	F
I	I	I	F
F	F	F	F

∨	T	I	F
T	T	T	T
I	T	I	I
F	T	I	F

→*K*	T	I	F
T	T	I	F
I	T	I	I
F	T	T	T

↔*K*	T	I	F
T	T	I	F
I	I	I	I
F	F	I	T

The difference between the two logics lies in how tautologies are defined. In K_3 only T is a designated truth value, while in P_3 both T and I are (a logical formula is considered a tautology if it evaluates to a designated truth value). In Kleene's logic I can be interpreted as being "underdetermined", being neither true nor false, while in Priest's logic I can be interpreted as being "overdetermined", being both true and false. K_3 does not have any tautologies, while P_3 has the same tautologies as classical two-valued logic.

Bochvar"s internal three-Valued Logic (also Known as Kleene"s weak Three-Valued Logic)

Another logic is Bochvar's "internal" three-valued logic (B_3^I) also called Kleene's weak three-valued logic. Except for negation and biconditional, its truth tables are all different from the above.

∧+	T	I	F
T	T	I	F
I	I	I	I
F	F	I	F

∨+	T	I	F
T	T	I	T
I	I	I	I
F	T	I	F

→+	T	I	F
T	T	I	F
I	I	I	I
F	T	I	T

The intermediate truth value in Bochvar's "internal" logic can be described as "contagious" because it propagates in a formula regardless of the value of any other variable.

Belnap logic (B_4)

Belnap's logic B_4 combines K_3 and P_3. The overdetermined truth value is here denoted as B and the underdetermined truth value as N.

$f_\neg$	
T	F
B	B
N	N
F	T

$f_\wedge$	T	B	N	F
T	T	B	N	F
B	B	B	F	F
N	N	F	N	F
F	F	F	F	F

$f_\vee$	T	B	N	F
T	T	T	T	T
B	T	B	T	B
N	T	T	N	N
F	T	B	N	F

Gödel logics G_k and G_∞

In 1932 Gödel defined a family G_k of many-valued logics, with finitely many truth values $0, \frac{1}{k-1}, \frac{2}{k-1}, \dots \frac{k-2}{k-1}, 1$, for example G_3 has the truth values $0, \frac{1}{2}, 1$ and G_4 has $0, \frac{1}{3}, \frac{2}{3}, 1$. In a similar manner he defined a logic with infinitely many truth values, G_∞, in which the truth values are all the real numbers in the interval [0,1]. The designated truth value in these logics is 1.

The conjunction $\wedge$ and the disjunction $\vee$ are defined respectively as the maximum and minimum of the operands:

- $u \wedge v := \min\{u, v\}$
- $u \vee v := \max\{u, v\}$

Negation $\neg_G$ and implication $\underset{G}{\rightarrow}$ are defined as follows:

$$\neg_G u = \begin{cases} 1, & \text{if } u = 0 \\ 0, & \text{if } u > 0 \end{cases}$$

$$u \underset{G}{\rightarrow} v = \begin{cases} 1, & \text{if } u \leq v \\ v, & \text{if } u > v \end{cases}$$

Gödel logics are completely axiomatisable, that is to say it is possible to define a logical calculus in which all tautologies are provable.

Łukasiewicz logics L_v and L_∞

Implication $\underset{L}{\rightarrow}$ and negation $\underset{L}{\neg}$ were defined by Jan Łukasiewicz through the following functions:

- $\underset{L}{\neg} u := 1 - u$
- $u \underset{L}{\rightarrow} v := min\{1, 1 - u + v\}$

At first Łukasiewicz used these definition in 1920 for his three-valued logic L_3 with truth values $0, \frac{1}{2}, 1$. In 1922 he developed a logic with infinitely many values L_∞, in which the truth values spanned the real numbers in the interval $[0,1]$. In both cases the designated truth value was 1.

By adopting truth values defined in the same way as for Gödel logics $0, \frac{1}{v-1}, \frac{2}{v-1}, \ldots, \frac{v-2}{v-1}, 1$, it is possible to create a finitely-valued family of logics L_v, the abovementioned L_∞ and the logic $L_{\aleph_0}$ in which the truth values are given by the rational numbers in the interval $[0,1]$. The set of tautologies in L_∞ and $L_{\aleph_0}$ is identical.

Product logic Π

In product logic we have truth values in the interval , $[0,1]$ a conjunction $\odot$ and an implication $\underset{\Pi}{\rightarrow}$, defined as follows

- $u \odot v := uv$
- $u \underset{\Pi}{\rightarrow} v := \begin{cases} 1, & \text{if } u \leq v \\ \frac{v}{u}, & \text{if } u > v \end{cases}$

Additionally there is a negative designated value $\overline{0}$ that denotes the concept of false. Through this value it is possible to define a negation $\underset{\Pi}{\neg}$ and an additional conjunction $\underset{\Pi}{\wedge}$ as follows:

- $\underset{\Pi}{\neg} u := u \underset{\Pi}{\rightarrow} \overline{0}$
- $u \underset{\Pi}{\wedge} v := u \odot (u \underset{\Pi}{\rightarrow} v)$

Post logics P_m

In 1921 Post defined a family of logics P_m with (as in L_v and G_k) the truth values $0, \frac{1}{m-1}, \frac{2}{m-1}, \ldots, \frac{m-2}{m-1}, 1$. Negation $\underset{P}{\neg}$ and disjunction $\underset{P}{\vee}$ are defined as follows:

- $\underset{P}{\neg} u := \begin{cases} 1, & \text{if } u = 0 \\ u - \frac{1}{m-1}, & \text{if } u \neq 0 \end{cases}$
- $u \underset{P}{\vee} v := max\{u, v\}$

Relation to Classical Logic

Logics are usually systems intended to codify rules for preserving some semantic property of propositions across transformations. In classical logic, this property is “truth.” In a valid argument, the truth of the derived proposition is guaranteed if the premises are jointly true, because the application of valid steps preserves the property. However, that property doesn”t have to be that of “truth”; instead, it can be some other concept.

Multi-valued logics are intended to preserve the property of designationhood (or being designated). Since there are more than two truth values, rules of inference may be intended to preserve more than just whichever corresponds (in the relevant sense) to truth. For example, in a three-valued logic, sometimes the two greatest truth-values (when they are represented as e.g. positive integers) are designated and the rules of inference preserve these values. Precisely, a valid argument will be such that the value of the premises taken jointly will always be less than or equal to the conclusion.

For example, the preserved property could be justification, the foundational concept of intuitionistic logic. Thus, a proposition is not true or false; instead, it is justified or flawed. A key difference between justification and truth, in this case, is that the law of excluded middle doesn't hold: a proposition that is not flawed is not necessarily justified; instead, it's only not proven that it's flawed. The key difference is the determinacy of the preserved property: One may prove that P is justified, that P is flawed, or be unable to prove either. A valid argument preserves justification across transformations, so a proposition derived from justified propositions is still justified. However, there are proofs in classical logic that depend upon the law of excluded middle; since that law is not usable under this scheme, there are propositions that cannot be proven that way.

Applications

Known applications of many-valued logic can be roughly classified into two groups. The first group uses many-valued logic domain to solve binary problems more efficiently. For example, a well-known approach to represent a multiple-output Boolean function is to treat its output part as a single many-valued variable and convert it to a single-output characteristic function. Other applications of many-valued logic include design of Programmable Logic Arrays (PLAs) with input decoders, optimization of finite state machines, testing, and verification.

The second group targets the design of electronic circuits which employ more than two discrete levels of signals, such as many-valued memories, arithmetic circuits, Field Programmable Gate Arrays (FPGA) etc. Many-valued circuits have a number of theoretical advantages over standard binary circuits. For example, the interconnect on and off chip can be reduced if signals in the circuit assume four or more levels rather than only two. In memory design, storing two instead of one bit of information per memory cell doubles the density of the memory in the same die size. Applications using arithmetic circuits often benefit from using alternatives to binary number systems. For example, residue and redundant number systems can reduce or eliminate the ripple-through carries which are involved in normal binary addition or subtraction, resulting in high-speed arithmetic operations. These number systems have a natural implementation using many-valued circuits. However, the practicality of these potential advantages heavily depends on the availability of circuit realizations, which must be compatible or competitive with present-day standard technologies. In addition to aiding in the design of electronic circuits, many-valued logic is used extensively to test circuits for faults and defects. Basically all known automatic test pattern generation (ATG) algorithms used for digital circuit testing require a simulator that can resolve 5-valued logic (0, 1, x, D, D").

THREE-VALUED LOGIC

In logic, a three-valued logic (also trinary logic, trivalent, ternary, or trilean, sometimes abbreviated 3VL) is any of several many-valued logic systems in which there are three truth values indicating true, false and some indeterminate third value. This is contrasted with the more commonly known bivalent logics (such as classical sentential or Boolean logic) which provide only for true and false. Conceptual form and basic ideas were initially created by Jan Łukasiewicz and C. I. Lewis. These were then re-formulated by Grigore Moisil in an axiomatic algebraic form, and also extended to n-valued logics in 1945.

Representation of Values

As with bivalent logic, truth values in ternary logic may be represented numerically using various representations of the ternary numeral system. A few of the more common examples are:

- in balanced ternary, each digit has one of 3 values: −1, 0, or +1; these values may also be simplified to −, 0, +, respectively;
- in the redundant binary representation, each digit can have a value of −1, 0, 0/1 (the value 0/1 has two different representations);
- in the ternary numeral system, each digit is a trit (trinary digit) having a value of: 0, 1, or 2;
- in the skew binary number system, only most-significant non-zero digit has a value 2, and the remaining digits have a value of 0 or 1;
- 1 for true, 2 for false, and 0 for unknown, unknowable/undecidable, irrelevant, or both;
- 0 for false, 1 for true, and a third non-integer "maybe" symbol such as ?, #, ½, or xy.

Inside a ternary computer, ternary values are represented by ternary signals.

This article mainly illustrates a system of ternary propositional logic using the truth values {false, unknown, true}, and extends conventional Boolean connectives to a trivalent context. Ternary predicate logics exist as well; these may have readings of the quantifier different from classical (binary) predicate logic and may include alternative quantifiers as well.

Logics

Where Boolean logic has $2^2 = 4$ unary operators, the addition of a third value in ternary logic leads to a total of $3^3 = 27$ distinct operators on a single input value. Similarly, where Boolean logic has $2^{22} = 16$ distinct binary operators (operators with 2 inputs), ternary logic has $3^{32} = 19{,}683$ such operators. Where we can easily name a significant fraction of the Boolean operators (not, and, or, nand, nor, exclusive or, equivalence, implication), it is unreasonable to attempt to name all but a small fraction of the possible ternary operators.

Kleene and Priest Logics

Below is a set of truth tables showing the logic operations for Kleene"s "strong logic of indeterminacy" and Priest"s "logic of paradox".

(F, false; U, unknown; T, true)

NOT(A)

A	¬A
F	T
U	U
T	F

AND(A, B)

A ∧ B		B		
		F	U	T
A	F	F	F	F
	U	F	U	U
	T	F	U	T

OR(A, B)

A ∨ B		B		
		F	U	T
A	F	F	U	T
	U	U	U	T
	T	T	T	T

(−1, false; 0, unknown; +1, true)

NEG(A)

A	¬A
−1	+1
0	0
+1	−1

MIN(A, B)

A ∧ B		B		
		−1	0	+1
A	−1	−1	−1	−1
	0	−1	0	0
	+1	−1	0	+1

MAX(A, B)

A ∨ B		B		
		−1	0	+1
A	−1	−1	0	+1
	0	0	0	+1
	+1	+1	+1	+1

In these truth tables, the unknown state can be thought of as neither true nor false in Kleene logic, or thought of as both true and false in Priest logic. The difference lies in the definition of tautologies. Where Kleene logic's only designated truth value is T, Priest logic's designated truth values are both T and U. In Kleene logic, the knowledge of whether any particular unknown state secretly represents true or false at any moment in time is not available. However, certain logical operations can yield an unambiguous result, even if they involve at least one unknown operand. For example, since true OR true equals true, and true OR false also equals true, one can infer that true OR unknown equals true, as well. In this example, since either bivalent state could be underlying the unknown state, but either state also yields the same result, a definitive true results in all three cases.

If numeric values, e.g. balanced ternary values, are assigned to false, unknown and true such that false is less than unknown and unknown is less than true, then A AND B AND C... = MIN(A, B, C ...) and A OR B OR C ... = MAX(A, B, C...).

Material implication for Kleene logic can be defined as:

$A \rightarrow B \stackrel{\text{def}}{=} \text{NOT}(A) \text{ OR } B$, and its truth table is

IMP$_K$(A, B), OR(¬A, B)

A → B		B		
		T	U	F
A	T	T	U	F
	U	T	U	U
	F	T	T	T

IMP$_K$(A, B), MAX(−A, B)

A → B		B		
		+1	0	−1
A	+1	+1	0	−1
	0	+1	0	0
	−1	+1	+1	+1

which differs from that for Łukasiewicz logic (described below).

Kleene logic has no tautologies (valid formulas) because whenever all of the atomic components of a well-formed formula are assigned the value Unknown, the formula itself must also have the value Unknown. (And the only designated truth value for Kleene logic is True.) However, the lack of valid formulas does not mean that it lacks valid arguments and/or inference rules. An argument is semantically valid in Kleene logic if, whenever (for any interpretation/model) all of its premises are True, the conclusion must also be True. (Note that the Logic of Paradox (LP) has the same truth tables as Kleene logic, but it has two designated truth values instead of one; these are: True and Both (the analogue of Unknown), so that LP does have tautologies but it has fewer valid inference rules.)

Łukasiewicz Logic

The Łukasiewicz Ł3 has the same tables for AND, OR, and NOT as the Kleene logic given above, but differs in its definition of implication in that "unknown implies unknown" is **true**. This section follows the presentation from Malinowski's chapter of the Handbook of the History of Logic, vol 8.

$IMP_Ł(A, B)$

A → B		B		
		T	U	F
A	T	T	U	F
	U	T	T	U
	F	T	T	T

$IMP_Ł(A, B)$

A → B		B		
		+1	0	−1
A	+1	+1	0	−1
	0	+1	+1	0
	−1	+1	+1	+1

In fact, using Łukasiewicz's implication and negation, the other usual connectives may be derived as:

- $A \vee B = (A \rightarrow B) \rightarrow B$
- $A \wedge B = \neg(\neg A \vee \neg B)$
- $A \leftrightarrow B = (A \rightarrow B) \wedge (B \rightarrow A)$

It's also possible to derive a few other useful unary operators (first derived by Tarski in 1921):

- $\mathbf{M}A = \neg A \rightarrow A$
- $\mathbf{L}A = \neg\mathbf{M}\neg A$
- $\mathbf{I}A = \mathbf{M}A \wedge \neg\mathbf{L}A$

They have the following truth tables:

A	MA
F	F
U	T
T	T

A	LA
F	F
U	F
T	T

A	IA
F	F
U	T
T	F

M is read as "it is not false that..." or in the (unsuccessful) Tarski–Łukasiewicz attempt to axiomatize modal logic using a three-valued logic, "it is possible that..." L is read "it is true that..." or "it is necessary that..." Finally I is read "it is unknown that..." or "it is contingent that..."

In Łukasiewicz"s Ł3 the designated value is True, meaning that only a proposition having this value everywhere is considered a tautology. For example, A → A and A ↔ A are tautologies in Ł3 and also in classical logic. Not all tautologies of classical logic lift to Ł3 "as is". For example, the law of excluded middle, A ∨ ¬A, and the law of non-contradiction, ¬(A ∧ ¬A) are not tautologies in Ł3. However, using the operator **I** defined above, it is possible to state tautologies that are their analogues:

- A ∨ **I**A ∨ ¬A (law of excluded fourth)
- ¬(A ∧ ¬**I**A ∧ ¬A) (extended contradiction principle).

Bochvar logic

Ternary Post logic

Modular algebras

Some 3VL modular algebras have been introduced more recently, motivated by circuit problems rather than philosophical issues:

- Cohn algebra
- Pradhan algebra
- Dubrova and Muzio algebra

Application in SQL

The database structural query language SQL implements ternary logic as a means of handling comparisons with NULL field content. The original intent of NULL in SQL was to represent missing data in a database, i.e. the assumption that an actual value exists, but that the value is not currently recorded in the database. SQL uses a common fragment of the Kleene K3 logic, restricted to AND, OR, and NOT tables. In SQL, the intermediate value is intended to be interpreted as UNKNOWN. Explicit comparisons with NULL, including that of another NULL yields UNKNOWN. However this choice of semantics is abandoned for some set operations, e.g. UNION or INTERSECT, where NULLs are treated as equal with each other. Critics assert that this inconsistency deprives SQL of intuitive semantics in its treatment of NULLs. The SQL standard defines an optional feature called F571, which adds some unary operators, among which IS UNKNOWN corresponding to the Łukasiewicz **I** in this article. The addition of IS UNKNOWN to the other operators of SQL's three-valued logic makes the SQL three-valued logic functionally complete, meaning its logical operators can express (in combination) any conceivable three-valued logical function.

FOUR-VALUED LOGIC

In logic, a **four-valued logic** is any logic with four truth values. Multiple such logics were invented to deal with various practical problems.

Applications

Four-valued logic taught on technical schools is used to model signal values in digital circuits: the four values are 1, 0, Z and X. 1 and 0 stand for boolean true and false, Z stands for high impedance or open circuit and X stands for don"t care (e.g., the value has no effect). This logic is itself a subset of 9-valued logic standard by the IEEE called IEEE 1164 and implemented, e.g., in VHDL"s std_logic.

Another four-valued logic is Belnap's relevance logic. Its possible values are true, false, both (true and false), and neither (true nor false). Belnap's logic is designed to cope with multiple information sources such that if only true is found then true is assigned, if only false is found then false is assigned, if some sources say true and others say false then both is assigned, and if no information is given by any information source then neither is assigned.

There is also a SAE J1939 standard, used for CAN data transmission in heavy road vehicles, which has four logical (boolean) values: False, True, Error Condition, and Not installed (represented by values 0–3). Error Condition means there is a technical problem obstacling data acquisition. The logics for that is for example True and Error Condition=Error Condition. Not installed is used for a feature which does not exist in this vehicle, and should be disregarded for logical calculation. On CAN, usually fixed data messages are sent containing many signal values each, so a signal representing a not-installed feature will be sent anyway.

TRUTH FUNCTION

In mathematical logic, a **truth function** is a function from a set of truth valuesto truth values. Classically the domain and range of a truth function are {truth, falsehood}, but they may have any number of truth values, including an infinity of these.

A logical connective is truth-functional if the truth-value of a compound sentence is a function of the truth-value of its sub-sentences. A class of connectives is truth-functional if each of its members is. For example, the connective "and" is truth-functional since a sentence like "Apples are fruits and carrots are vegetables" is true if, and only if each of its sub-sentences "apples are fruits" and "carrots are vegetables" is true, and it is false otherwise. Some connectives of a natural language, such as English, are not truth-functional.

Connectives of the form "x believes that ..." are typical examples of connectives that are not truth-functional. If e.g. Mary mistakenly believes that Al Gore was President of the USA on April 20, 2000, but she does not believe that the moon is made of green cheese, then the sentence

"Mary believes that Al Gore was President of the USA on April 20, 2000"

is true while

"Mary believes that the moon is made of green cheese"

is false. In both cases, each component sentence (i.e. "Al Gore was president of the USA on April 20, 2000" and "the moon is made of green cheese") is false, but each compound sentence formed by prefixing the phrase "Mary believes that" differs in truth-value. That is, the truth-value of a sentence of the form "Mary believes that..." is not determined solely by the truth-value of its component sentence, and hence the (unary) connective (or simply operator since it is unary) is non-truth-functional.

The class of classical logic connectives (e.g. &, →) used in the construction of formulas is truth-functional. Their values for various truth-values as argument are usually given by truth tables. Truth-functional propositional calculus is a formal system whose formulas may be interpreted as either true or false.

Table of Binary Truth Functions

In two-valued logic, there are sixteen possible truth functions, also called Boolean functions, of two inputs P and Q. Any of these functions corresponds to a truth table of a certain logical connective in classical logic, including several degenerate cases such as a function not depending on one or both of its arguments. Truth and falsehood is denoted as 1 and 0 in the following truth tables, respectively, for sake of brevity.

Contradiction/False			
Notation	Equivalent formulas	Truth table	Venn diagram
⊥ "bottom"	$P \wedge \neg P$ Opq	Q: 0, 1 P 0: 0, 0 P 1: 0, 0	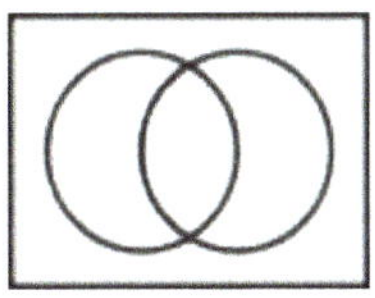

Tautology/True			
Notation	Equivalent formulas	Truth table	Venn diagram
⊤ "top"	$P \vee \neg P$ V*pq*	Q: 0 1 P 0: 1 1 P 1: 1 1	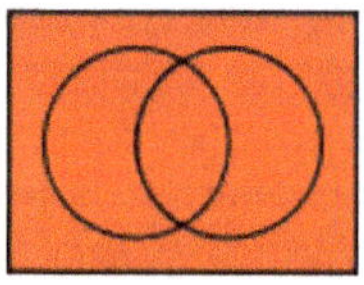

Proposition *P*			
Notation	Equivalent formulas	Truth table	Venn diagram
P	*p* I*pq*	Q: 0 1 P 0: 0 0 P 1: 1 1	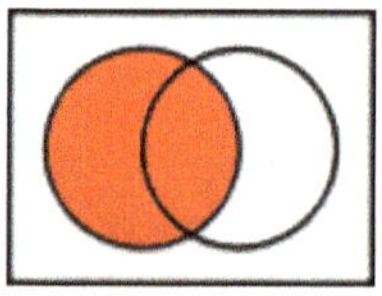

Negation of *Q*			
Notation	Equivalent formulas	Truth table	Venn diagram
$\neg Q$ ~*Q*	N*q* G*pq*	Q: 0 1 P 0: 1 0 P 1: 1 0	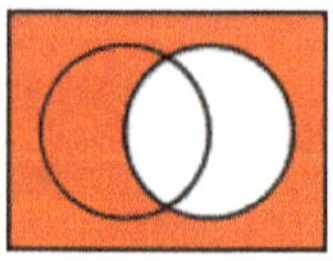

Conjunction			
Notation	Equivalent formulas	Truth table	Venn diagram
$P \land Q$ $P \,\&\, Q$ $P \cdot Q$ P AND Q	$P \nrightarrow \neg Q$ $\neg P \nleftarrow Q$ $\neg P \downarrow \neg Q$ Kpq	Q: 0, 1 P = 0: 0, 0 P = 1: 0, 1	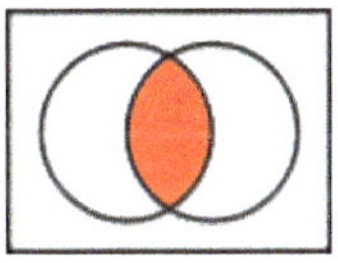

Alternative denial			
Notation	Equivalent formulas	Truth table	Venn diagram
$P \uparrow Q$ $P \mid Q$ P NAND Q	$P \rightarrow \neg Q$ $\neg P \leftarrow Q$ $\neg P \lor \neg Q$ Dpq	Q: 0, 1 P = 0: 1, 1 P = 1: 1, 0	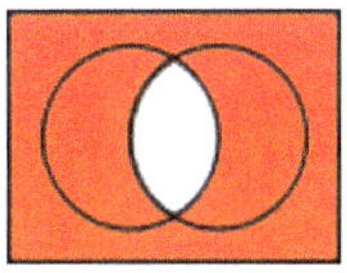

Disjunction			
Notation	Equivalent formulas	Truth table	Venn diagram
$P \lor Q$ P OR Q	$P \leftarrow \neg Q$ $\neg P \rightarrow Q$ $\neg P \uparrow \neg Q$ $\neg(\neg P \land \neg Q)$ Apq	Q: 0, 1 P = 0: 0, 1 P = 1: 1, 1	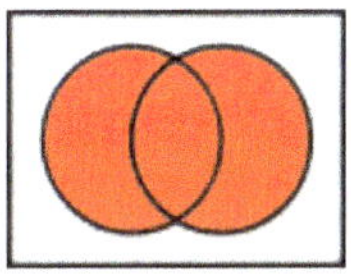

Joint denial			
Notation	Equivalent formulas	Truth table	Venn diagram
$P \downarrow Q$ P NOR Q	$P \nleftarrow \neg Q$ $\neg P \nrightarrow Q$ $\neg P \land \neg Q$ Xpq	Q: 0, 1 P = 0: 1, 0 P = 1: 0, 0	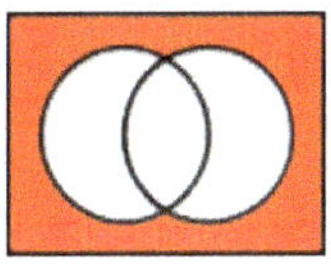

Material nonimplication			
Notation	Equivalent formulas	Truth table	Venn diagram
$P \nrightarrow Q$ $P \not\supset Q$	$P \wedge \neg Q$ $\neg P \downarrow Q$ $\neg P \nleftarrow \neg Q$ Lpq	see below	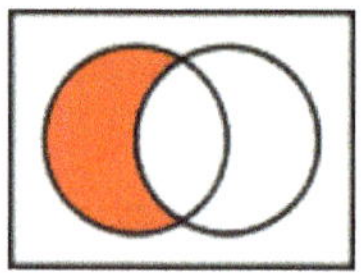

		Q = 0	Q = 1
P	0	0	0
P	1	1	0

Material implication			
Notation	Equivalent formulas	Truth table	Venn diagram
$P \rightarrow Q$ $P \supset Q$	$P \uparrow \neg Q$ $\neg P \vee Q$ $\neg P \leftarrow \neg Q$ Cpq	see below	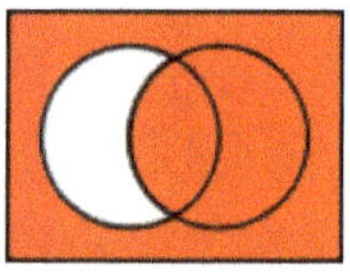

		Q = 0	Q = 1
P	0	1	1
P	1	0	1

Converse nonimplication			
Notation	Equivalent formulas	Truth table	Venn diagram
$P \nleftarrow Q$ $P \not\subset Q$	$P \downarrow \neg Q$ $\neg P \wedge Q$ $\neg P \nrightarrow \neg Q$ Mpq	see below	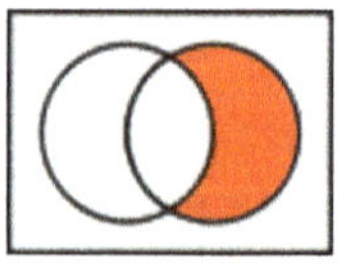

		Q = 0	Q = 1
P	0	0	1
P	1	0	0

Converse implication			
Notation	Equivalent formulas	Truth table	Venn diagram
$P \leftarrow Q$ $P \subset Q$	$P \vee \neg Q$ $\neg P \uparrow Q$ $\neg P \rightarrow \neg Q$ Bpq	Q: 0, 1 P = 0: 1, 0 P = 1: 1, 1	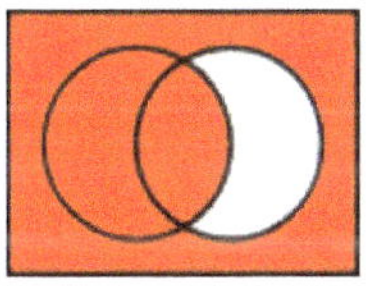

Exclusive disjunction			
Notation	Equivalent formulas	Truth table	Venn diagram
$P \not\leftrightarrow Q$ $P \not\equiv Q$ $P \oplus Q$ P XOR Q	$P \leftrightarrow \neg Q$ $\neg P \leftrightarrow Q$ $\neg P \not\leftrightarrow \neg Q$ Jpq	Q: 0, 1 P = 0: 0, 1 P = 1: 1, 0	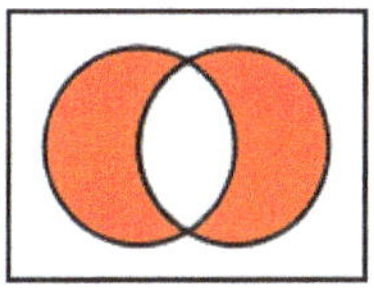

Biconditional			
Notation	Equivalent formulas	Truth table	Venn diagram
$P \leftrightarrow Q$ $P \equiv Q$ P XNOR Q P IFF Q	$P \not\leftrightarrow \neg Q$ $\neg P \not\leftrightarrow Q$ $\neg P \leftrightarrow \neg Q$ Epq	Q: 0, 1 P = 0: 1, 0 P = 1: 0, 1	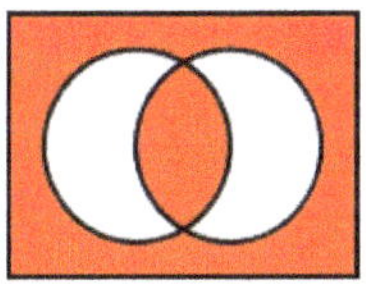

Functional Completeness

Because a function may be expressed as a composition, a truth-functional logical calculus does not need to have dedicated symbols for all of the above-mentioned functions to be functionally complete. This is expressed in a propositional calculus as logical equivalence of certain compound statements. For example, classical logic has $\neg P \vee Q$ equivalent to $P \rightarrow Q$. The conditional operator "$\rightarrow$" is therefore not necessary for a classical-based logical system if "$\neg$" (not) and "$\vee$" (or) are already in use.

A minimal set of operators that can express every statement expressible in the propositional calculus is called a minimal functionally complete set. A minimally complete set of operators is achieved by NAND alone $\{\uparrow\}$ and NOR alone $\{\downarrow\}$.

The following are the minimal functionally complete sets of operators whose arities do not exceed 2:

One element

$\{\uparrow\}$, $\{\downarrow\}$.

Two elements

$\{\vee, \neg\}$, $\{\wedge, \neg\}$, $\{\rightarrow, \neg\}$, $\{\leftarrow, \neg\}$, $\{\rightarrow, \bot\}$, $\{\leftarrow, \bot\}$, $\{\rightarrow, \nleftrightarrow\}$, $\{\leftarrow, \nleftrightarrow\}$, $\{\rightarrow, \nrightarrow\}$, $\{\rightarrow, \nleftarrow\}$, $\{\leftarrow, \nrightarrow\}$, $\{\leftarrow, \nleftarrow\}$, $\{\nrightarrow, \neg\}$, $\{\nleftarrow, \neg\}$, $\{\nrightarrow, \top\}$, $\{\nleftarrow, \top\}$, $\{\nrightarrow, \leftrightarrow\}$, $\{\nleftarrow, \leftrightarrow\}$.

Three elements

$\{\vee, \leftrightarrow, \bot\}$, $\{\vee, \leftrightarrow, \nleftrightarrow\}$, $\{\vee, \nleftrightarrow, \top\}$, $\{\wedge, \leftrightarrow, \bot\}$, $\{\wedge, \leftrightarrow, \nleftrightarrow\}$, $\{\wedge, \nleftrightarrow, \top\}$.

Algebraic Properties

Some truth functions possess properties which may be expressed in the theorems containing the corresponding connective. Some of those properties that a binary truth function (or a corresponding logical connective) may have are:

- **Associativity**: Within an expression containing two or more of the same associative connectives in a row, the order of the operations does not matter as long as the sequence of the operands is not changed.
- **Commutativity**: The operands of the connective may be swapped without affecting the truth-value of the expression.
- **Distributivity**: A connective denoted by · distributes over another connective denoted by +, if $a \cdot (b + c) = (a \cdot b) + (a \cdot c)$ for all operands a, b, c.
- **Idempotence**: Whenever the operands of the operation are the same, the connective gives the operand as the result. In other words, the operation is both truth-preserving and falsehood-preserving (see below).
- **Absorption**: A pair of connectives $\wedge, \vee$ satisfies the absorption law if $a \wedge (a \vee b) = a$ for all operands a, b.

A set of truth functions is functionally complete if and only if for each of the following five properties it contains at least one member lacking it:

- **monotonic**: If $f(a_1, ..., a_n) \leq f(b_1, ..., b_n)$ for all $a_1, ..., a_n, b_1, ..., b_n \in \{0,1\}$ such that $a_1 \leq b_1$, $a_2 \leq b_2$, ..., $a_n \leq b_n$. E.g., $\vee, \wedge, \top, \bot$.

- **affine**: Each variable always makes a difference in the truth-value of the operation or it never makes a difference. E.g., $\neg, \leftrightarrow, \nleftrightarrow, \top, \bot$.
- **self dual**: To read the truth-value assignments for the operation from top to bottom on its truth table is the same as taking the complement of reading it from bottom to top; in other words, $f(\neg a_1, ..., \neg a_n) = \neg f(a_1, ..., a_n)$. E.g., $\neg$.
- **truth-preserving**: The interpretation under which all variables are assigned a truth value of 'true' produces a truth value of 'true' as a result of these operations. E.g., $\vee, \wedge, \top, \rightarrow, \leftrightarrow, \subset$. (see validity)
- **falsehood-preserving**: The interpretation under which all variables are assigned a truth value of 'false' produces a truth value of 'false' as a result of these operations. E.g $\vee, \wedge, \nleftrightarrow, \bot, \not\subset, \not\supset$. (see validity)

Arity

A concrete function may be also referred to as an operator. In two-valued logic there are 2 nullary operators (constants), 4 unary operators, 16 binary operators, 256 ternary operators, and 2^{2^n} n-ary operators. In three-valued logic there are 3 nullary operators (constants), 27 unary operators, 19683 binary operators, 7625597484987 ternary operators, and 3^{3^n} n-ary operators. In k-valued logic, there are k nullary operators, k^k unary operators, k^{k^2}

binary operators, k^{k^3} ternary operators, and k^{k^n} n-ary operators. An n-ary operator in k-valued logic is a function from $\mathbb{Z}_k^n \to \mathbb{Z}_k$. Therefore, the number of such operators is $|\mathbb{Z}_k|^{|\mathbb{Z}_k^n|} = k^{k^n}$, which is how the above numbers were derived.

However, some of the operators of a particular arity are actually degenerate forms that perform a lower-arity operation on some of the inputs and ignores the rest of the inputs. Out of the 256 ternary boolean operators cited above, $\binom{3}{2} \cdot 16 - \binom{3}{1} \cdot 4 + \binom{3}{0} \cdot 2$ of them are such degenerate forms of binary or lower-arity operators, using the inclusion-exclusion principle. The ternary operator $f(x, y, z) = \neg x$ is one such operator which is actually a unary operator applied to one input, and ignoring the other two inputs.

"Not" is a unary operator, it takes a single term (¬P). The rest are binary operators, taking two terms to make a compound statement (P ∧ Q, P ∨ Q, P → Q, P ↔ Q).

The set of logical operators Ω may be partitioned into disjoint subsets as follows:

$$\Omega = \Omega_0 \cup \Omega_1 \cup \ldots \cup \Omega_j \cup \ldots \cup \Omega_m .$$

In this partition, Ω_j is the set of operator symbols of arity j.

In the more familiar propositional calculi, Ω is typically partitioned as follows:

nullary operators: $\Omega_0 = \{\perp, \top\}$

unary operators: $\Omega_1 = \{\neg\}$

binary operators: $\Omega_2 \subseteq \{\wedge, \vee, \rightarrow, \leftrightarrow\}$

Principle of Compositionality

Instead of using truth tables, logical connective symbols can be interpreted by means of an interpretation function and a functionally complete set of truth-functions (Gamut 1991), as detailed by the principle of compositionality of meaning. Let I be an interpretation function, let Φ, Ψ be any two sentences and let the truth function f_{nand} be defined as:

- $f_{nand}(T,T)=F$; $f_{nand}(T,F)=f_{nand}(F,T)=f_{nand}(F,F)=T$

Then, for convenience, f_{not}, f_{or} f_{and} and so on are defined by means of f_{nand}:

- $f_{not}(x)=f_{nand}(x,x)$
- $f_{or}(x,y)= f_{nand}(f_{not}(x), f_{not}(y))$
- $f_{and}(x,y)=f_{not}(f_{nand}(x,y))$

or, alternatively f_{not}, f_{or} f_{and} and so on are defined directly:

- $f_{not}(T)=F$; $f_{not}(F)=T$;
- $f_{or}(T,T)=f_{or}(T,F)=f_{or}(F,T)=T$;$f_{or}(F,F)=F$
- $f_{and}(T,T)=T$; $f_{and}(T,F)=f_{and}(F,T)=f_{and}(F,F)=F$

Then

- $I(\sim)=I(\neg)=f_{not}$
- $I(\&)=I(\wedge)=f_{and}$
- $I(\text{v})=I(\vee)=f_{or}$
- $I(\sim\Phi)=I(\neg\Phi)=I(\neg)(I(\Phi))=f_{not}(I(\Phi))$
- $I(\Phi\wedge\Psi) = I(\wedge)(I(\Phi), I(\Psi))=f_{and}(I(\Phi), I(\Psi))$

etc.

Thus if S is a sentence that is a string of symbols consisting of logical symbols v_1... v_n representing logical connectives, and non-logical symbols c_1...c_n, then if and only if $I(v_1)$...$I(v_n)$ have been provided interpreting v_1 to v_n by means of f_{nand} (or any other set of functional complete truth-functions) then the truth-value of {\displaystyle I(s)} is determined entirely by the truth-values of c_1...c_n, i.e. of $I(c_1)$...$I(c_n)$. In other words, as expected and required, S is true or false only under an interpretation of all its non-logical symbols.

Computer Science

Logical operators are implemented as logic gates in digital circuits. Practically all digital circuits (the major exception is DRAM) are built up from NAND, NOR, NOT, and transmission gates. NAND and NOR gates with 3 or more inputs rather than the usual 2 inputs are fairly common, although they are logically equivalent to a cascade of

2-input gates. All other operators are implemented by breaking them down into a logically equivalent combination of 2 or more of the above logic gates.

The "logical equivalence" of "NAND alone", "NOR alone", and "NOT and AND" is similar to Turing equivalence.

The fact that all truth functions can be expressed with NOR alone is demonstrated by the Apollo guidance computer.

REFERENCES

1. Hurley, Patrick. A Concise Introduction to Logic, 9th edition. (2006).
2. (Gottwald 2005, p. 19)
3. (Bergmann 2008, p. 80)
4. Gödel, Kurt (1932). "Zum intuitionistischen Aussagenkalkül". Anzeiger der Akademie der Wissenschaften in Wien (69): 65f.
5. Kreiser, Lothar; Gottwald, Siegfried; Stelzner, Werner (1990). Nichtklassische Logik. Eine Einführung. Berlin: Akademie-Verlag. pp. 41ff — 45ff. ISBN 978-3-05-000274-3.
6. Hajek, Petr: Fuzzy Logic. In: Edward N. Zalta: The Stanford Encyclopedia of Philosophy, Spring 2009. ()
7. Dubrova, Elena (2002). Multiple-Valued Logic Synthesis and Optimization, in Hassoun S. and Sasao T., editors, Logic Synthesis and Verification, Kluwer Academic Publishers, pp. 89-114
8. Meher, Pramod Kumar; Valls, Javier; Juang, Tso-Bing; Sridharan, K.; Maharatna, Koushik (2008-08-22). "50 Years of CORDIC: Algorithms, Architectures and Applications" (PDF). IEEE Transactions on Circuits & Systems-I: Regular Papers (published 2009-09-09). **56** (9): 1893–1907. Retrieved 2016-01-03.
9. http://www.informatik.uni-trier.de/~ley/db/conf/ismvl/index.html
10. http://www.oldcitypublishing.com/MVLSC/MVLSC.html
11. Aristotle, "The Categories", Harold P. Cooke (trans.), pp. 1–109 in Aristotle, Volume 1, Loeb Classical Library, William Heinemann, London, UK, 1938.
12. Aristotle, "On Interpretation", Harold P. Cooke (trans.), pp. 111–179 in Aristotle, Volume 1, Loeb Classical Library, William Heinemann, London, UK, 1938.
13. Aristotle, "Prior Analytics", Hugh Tredennick (trans.), pp. 181–531 in Aristotle, Volume 1, Loeb Classical Library, William Heinemann, London, UK, 1938.
14. Aristotle, "On the Soul" (De Anima), W. S. Hett (trans.), pp. 1–203 in Aristotle, Volume 8, Loeb Classical Library, William Heinemann, London, UK, 1936.
15. Audi, Robert (ed., 1999), The Cambridge Dictionary of Philosophy, Cambridge University Press, Cambridge, UK, 1995. 2nd edition, 1999. Cited as CDP.
16. Baldwin, James Mark (ed., 1901–1905), Dictionary of Philosophy and Psychology, 3 volumes in 4, Macmillan, New York, NY.
17. Baylis, Charles A. (1962), "Truth", pp. 321–322 in Dagobert D. Runes (ed.), Dictionary of Philosophy, Littlefield, Adams, and Company, Totowa, NJ.
18. Benjamin, A. Cornelius (1962), "Coherence Theory of Truth", p. 58 in Dagobert D. Runes (ed.), Dictionary of Philosophy, Littlefield, Adams, and Company, Totowa, NJ.

19. Blackburn, Simon, and Simmons, Keith (eds., 1999), Truth, Oxford University Press, Oxford, UK. Includes papers by James, Ramsey, Russell, Tarski, and more recent work.

20. Chandrasekhar, Subrahmanyan (1987), Truth and Beauty. Aesthetics and Motivations in Science, University of Chicago Press, Chicago, IL.

21. Chang, C.C., and Keisler, H.J., Model Theory, North-Holland, Amsterdam, Netherlands, 1973.

22. Chomsky, Noam (1995), The Minimalist Program, MIT Press, Cambridge, MA.

23. Church, Alonzo (1962a), "Name Relation, or Meaning Relation", p. 204 in Dagobert D. Runes (ed.), Dictionary of Philosophy, Littlefield, Adams, and Company, Totowa, NJ.

24. Church, Alonzo (1962b), "Truth, Semantical", p. 322 in Dagobert D. Runes (ed.), Dictionary of Philosophy, Littlefield, Adams, and Company, Totowa, NJ.

25. Clifford, W.K. (1877), "The Ethics of Belief and Other Essays". (Prometheus Books, 1999), infidels.org

26. Dewey, John (1900–1901), Lectures on Ethics 1900–1901, Donald F. Koch (ed.), Southern Illinois University Press, Carbondale and Edwardsville, IL.

27. Dewey, John (1932), Theory of the Moral Life, Part 2 of John Dewey and James H. Tufts, Ethics, Henry Holt and Company, New York, NY, 1908. 2nd edition, Holt, Rinehart, and Winston, 1932. Reprinted, Arnold Isenberg (ed.), Victor Kestenbaum (pref.), Irvingtion Publishers, New York, NY, 1980.

28. Dewey, John (1938), Logic: The Theory of Inquiry (1938),Holt and Company, New York, NY. Reprinted, John Dewey, The Later Works, 1925–1953, Volume 12: 1938, Jo Ann Boydston (ed.), Southern Illinois University Press, Carbondale and Edwardsville, IL, 1986.

29. Field, Hartry (2001), Truth and the Absence of Fact, Oxford University Press, Oxford, UK.

30. Foucault, Michel (1997), Essential Works of Foucault, 1954–1984, Volume 1, Ethics: Subjectivity and Truth, Paul Rabinow (ed.), Robert Hurley et al. (trans.), The New Press, New York, NY.

31. Garfield, Jay L., and Kiteley, Murray (1991), Meaning and Truth: The Essential Readings in Modern Semantics, Paragon House, New York, NY.

32. Gupta, Anil (2001), "Truth", in Lou Goble (ed.), The Blackwell Guide to Philosophical Logic, Blackwell Publishers, Oxford, UK.

33. Gupta, Anil and Belnap, Nuel. (1993). The Revision Theory of Truth. MIT Press.

34. Haack, Susan (1993), Evidence and Inquiry: Towards Reconstruction in Epistemology, Blackwell Publishers, Oxford, UK.

35. Habermas, Jürgen (1976), "What Is Universal Pragmatics?", 1st published, "Was heißt Universalpragmatik?", Sprachpragmatik und Philosophie, Karl-Otto Apel (ed.), Suhrkamp Verlag, Frankfurt am Main. Reprinted, pp. 1–68 in Jürgen Habermas, Communication and the Evolution of Society, Thomas McCarthy (trans.), Beacon Press, Boston, MA, 1979.

36. Habermas, Jürgen (1990), Moral Consciousness and Communicative Action, Christian Lenhardt and Shierry Weber Nicholsen (trans.), Thomas McCarthy (intro.), MIT Press, Cambridge, MA.

37. Habermas, Jürgen (2003), Truth and Justification, Barbara Fultner (trans.), MIT Press, Cambridge, MA.

38. Hegel, Georg, (1977), Phenomenology of Spirit, Oxford University Press, Oxford, UK, ISBN 0-19-824597-1.

39. Horwich, Paul, (1988), Truth, 2nd edition, Oxford University Press, Oxford, UK.
40. James, William (1904), A World of Pure Experience.
41. James, William (1907), Pragmatism, A New Name for Some Old Ways of Thinking, Popular Lectures on Philosophy, Longmans, Green, and Company, New York, NY.
42. James, William (1909), The Meaning of Truth, A Sequel to ‹Pragmatism", Longmans, Green, and Company, New York, NY.
43. James, William (1912), Essays in Radical Empiricism. Cf. Chapt. 3, "The Thing and its Relations", pp. 92–122.
44. James, William (2014), William James on Habit, Will, Truth, and the Meaning of Life. James Sloan Allen (ed.), Frederic C. Beil, Publisher, Savannah, GA.
45. Kant, Immanuel (1800), Introduction to Logic. Reprinted, Thomas Kingsmill Abbott (trans.), Dennis Sweet (intro.), Barnes and Noble, New York, NY, 2005.
46. Kirkham, Richard L. (1992), Theories of Truth: A Critical Introduction, MIT Press, Cambridge, MA.
47. Kneale, W., and Kneale, M. (1962), The Development of Logic, Oxford University Press, London, UK, 1962. Reprinted with corrections, 1975.
48. Kreitler, Hans, and Kreitler, Shulamith (1972), Psychology of the Arts, Duke University Press, Durham, NC.
49. Le Morvan, Pierre (2004), "Ramsey on Truth and Truth on Ramsey", British Journal for the History of Philosophy, 12 (4) 2004, 705–718, PDF.
50. Peirce, C.S., Bibliography.
51. Peirce, C.S., Collected Papers of Charles Sanders Peirce, vols. 1–6, Charles Hartshorne and Paul Weiss (eds.), vols. 7–8, Arthur W. Burks (ed.), Harvard University Press, Cambridge, MA, 1931–1935, 1958. Cited as CP vol.para.
52. Peirce, C.S. (1877), "The Fixation of Belief", Popular Science Monthly 12 (1877), 1–15. Reprinted (CP 5.358–387), (CE 3, 242–257), (EP 1, 109–123). Eprint.
53. Peirce, C.S. (1901), "Truth and Falsity and Error" (in part), pp. 718–720 in J.M. Baldwin (ed.), Dictionary of Philosophy and Psychology, vol. 2. Reprinted, CP 5.565–573.
54. Polanyi, Michael (1966), The Tacit Dimension, Doubleday and Company, Garden City, NY.
55. Quine, W.V. (1956), "Quantifiers and Propositional Attitudes", Journal of Philosophy 53 (1956). Reprinted, pp. 185–196 in Quine (1976), Ways of Paradox.
56. Quine, W.V. (1976), The Ways of Paradox, and Other Essays, 1st edition, 1966. Revised and enlarged edition, Harvard University Press, Cambridge, MA, 1976.
57. Quine, W.V. (1980 a), From a Logical Point of View, Logico-Philosophical Essays, 2nd edition, Harvard University Press, Cambridge, MA.
58. Quine, W.V. (1980 b), "Reference and Modality", pp. 139–159 in Quine (1980 a), From a Logical Point of View.
59. Rajchman, John, and West, Cornel (ed., 1985), Post-Analytic Philosophy, Columbia University Press, New York, NY.
60. Ramsey, F.P. (1927), "Facts and Propositions", Aristotelian Society Supplementary Volume 7, 153–170. Reprinted, pp. 34–51 in F.P. Ramsey, Philosophical Papers, David Hugh Mellor (ed.), Cambridge University Press, Cambridge, UK, 1990.
61. Ramsey, F.P. (1990), Philosophical Papers, David Hugh Mellor (ed.), Cambridge University Press, Cambridge, UK.

62. Rawls, John (2000), Lectures on the History of Moral Philosophy, Barbara Herman (ed.), Harvard University Press, Cambridge, MA.

63. Rorty, R. (1979), Philosophy and the Mirror of Nature, Princeton University Press, Princeton, NJ.

64. Russell, Bertrand (1912), The Problems of Philosophy, 1st published 1912. Reprinted, Galaxy Book, Oxford University Press, New York, NY, 1959. Reprinted, Prometheus Books, Buffalo, NY, 1988.

65. Russell, Bertrand (1918), "The Philosophy of Logical Atomism", The Monist, 1918. Reprinted, pp. 177–281 in Logic and Knowledge: Essays 1901–1950, Robert Charles Marsh (ed.), Unwin Hyman, London, UK, 1956. Reprinted, pp. 35–155 in The Philosophy of Logical Atomism, David Pears (ed.), Open Court, La Salle, IL, 1985.

66. Russell, Bertrand (1956), Logic and Knowledge: Essays 1901–1950, Robert Charles Marsh (ed.), Unwin Hyman, London, UK, 1956. Reprinted, Routledge, London, UK, 1992.

67. Russell, Bertrand (1985), The Philosophy of Logical Atomism, David Pears (ed.), Open Court, La Salle, IL.

68. Schopenhauer, Arthur, (1974), On the Fourfold Root of the Principle of Sufficient Reason, Open Court, La Salle, IL, ISBN 0-87548-187-6.

69. Smart, Ninian (1969), The Religious Experience of Mankind, Charles Scribner"s Sons, New York, NY.

70. Tarski, A., Logic, Semantics, Metamathematics: Papers from 1923 to 1938, J.H. Woodger (trans.), Oxford University Press, Oxford, UK, 1956. 2nd edition, John Corcoran (ed.), Hackett Publishing, Indianapolis, IN, 1983.

71. Wallace, Anthony F.C. (1966), Religion: An Anthropological View, Random House, New York, NY.

72. Williamson, T. 1994. Vagueness London: Routledge.he history of the problem of vagueness is traced, from the first Sorites Paradox to contemporary attempts to deal with higher-order vagueness such as many-valued logic, supervaluationism, and fuzzy logic. Technicalities are kept to a minimum to favour a clear account, extremely useful to both students and researchers.

73. Machina, K.F. 1976. "Truth, belief and vagueness", in Journal of Philosophical Logic Vol. 5. pp. 47-78

74. Edgington, D. (1997). Keefe, R.; Smith, P., eds. Vagueness by degrees (PDF). MIT Press. pp. 294–316.

75. Kit Fine, The Limits of Abstraction (2002)

76. Sorensen, R.A. 1988. Blindspots. Oxford: Clarendon Press.

77. Sorensen, Roy (2001). Vagueness and Contradiction. Oxford University Press.

78. Rescher, N. 2009. Unknowability. Lexington Books.Uses vagrant predicates to elucidate the problem.

79. Evans, G. (1978). "Can There Be Vague Objects?". Analysis. **38**: 208–. doi:10.1093/analys/38.4.208.

80. Van Inwagen, Peter. 1990 Material Beings. Ithaca, NY: Cornell University Press.

81. Parsons, Terence. 2000. Indeterminate Identity - Metaphysics and Semantics Oxford: Clarendon Press.

CHAPTER 6

Interval Arithmetic

Chapter Contents

- **Introduction**
- **Complex Interval Arithmetic**
- **Interval Methods**
- **Application**
- **History**
- **Implementations**
- **Conferences And Workshop**

Interval arithmetic, interval mathematics, interval analysis, or interval computation, is a method developed by mathematicians since the 1950s and 1960s, as an approach to putting bounds on rounding errors and measurement errors in mathematical computation and thus developing numerical methods that yield reliable results. Very simply put, it represents each value as a range of possibilities. For example, instead of estimating the height of someone using standard arithmetic as 2.0 metres, using interval arithmetic we might be certain that that person is somewhere between 1.97 and 2.03 metres.

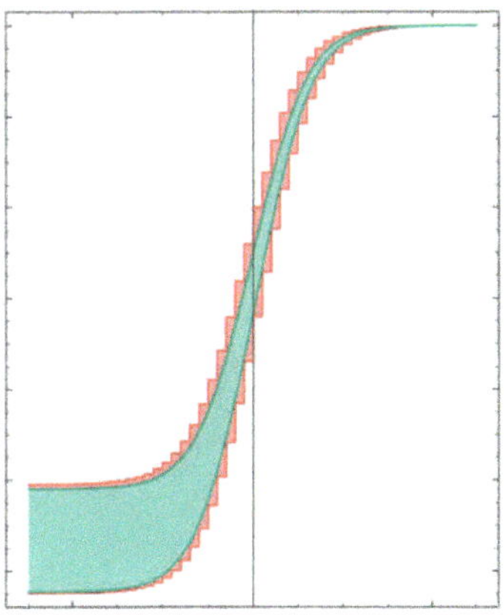

Tolerance function (turquoise) and interval-valued approximation (red)

This concept is suitable for a variety of purposes. The most common use is to keep track of and handle rounding errors directly during the calculation and of uncertainties in the knowledge of the exact values of physical and technical parameters. The latter often arise from measurement errors and tolerances for components or due to limits on computational accuracy. Interval arithmetic also helps find reliable and guaranteed solutions to equations and optimization problems.

Mathematically, instead of working with an uncertain real x we work with the two ends of the interval [a,b] that contains x. In interval arithmetic, any variable x lies between a and b, or could be one of them. A function f when applied to x is also uncertain. In interval arithmetic f produces an interval [c,d] that is all the possible values for $f(x)$ for all $x \in [a, b]$.

INTRODUCTION

The main focus of interval arithmetic is the simplest way to calculate upper and lower endpoints for the range of values of a function in one or more variables. These endpoints are not necessarily the supremum or infimum, since the precise calculation of those values can be difficult or impossible.

Treatment is typically limited to real intervals, so quantities of form

$$[a, b] = \{x \in \mathbb{R} \,|\, a \leq x \leq b\},$$

where $a = -\infty$ and $b = \infty$ are allowed; with one of them infinite we would have an unbounded interval, while with both infinite we would have the extended real number line.

As with traditional calculations with real numbers, simple arithmetic operations and functions on elementary intervals must first be defined. More complicated functions can be calculated from these basic elements.

Example

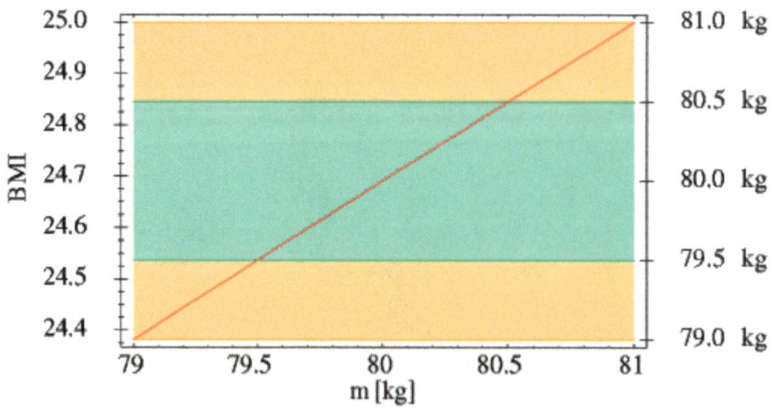

Body Mass Index for a person 1.80 m tall in relation to body weight *m* (in kilograms)

Take as an example the calculation of body mass index (BMI). The BMI is the body weight in kilograms divided by the square of height in metres. Measuring the mass with bathroom scales may have an accuracy of one kilogram. We do not know intermediate values – about 79.6 kg or 80.3 kg – but information rounded to the nearest whole number. It is unlikely that when the scale reads 80 kg, someone really weighs exactly 80.0 kg. In normal rounding to the nearest value, the scales showing 80 kg indicates a weight between 79.5 kg and 80.5 kg. The relevant range is that of all real numbers that are greater than or equal to 79.5, while less than or equal to 80.5, or in other words the interval [79.5,80.5].

For a man who weighs 80 kg and is 1.80 m tall, the BMI is about 24.7. With a weight of 79.5 kg and the same height the value is 24.5, while 80.5 kilograms gives almost 24.9. So the actual BMI is in the range [24.5,24.9]. The error in this case does not affect the conclusion (normal weight), but this is not always the position. For example, weight fluctuates in the course of a day so that the BMI can vary between 24 (normal weight) and 25 (overweight). Without detailed analysis it is not possible to always exclude questions as to whether an error ultimately is large enough to have significant influence.

Interval arithmetic states the range of possible outcomes explicitly. Simply put, results are no longer stated as numbers, but as intervals that represent imprecise values. The size of the intervals are similar to error bars to a metric in expressing the extent of uncertainty. Simple arithmetic operations, such as basic arithmetic and trigonometric functions, enable the calculation of outer limits of intervals.

Simple Arithmetic

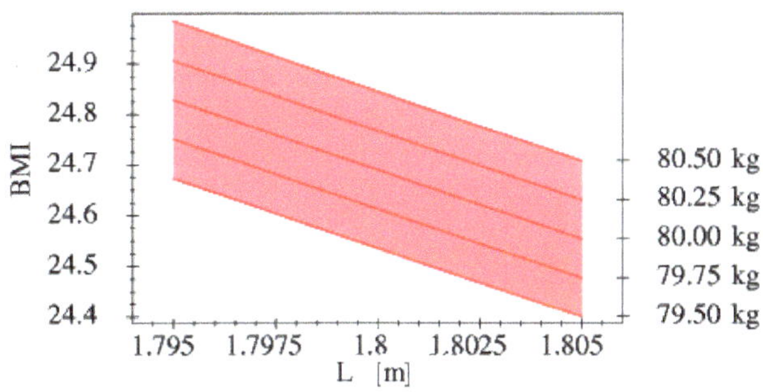

Body mass index for different weights in relation to height L (in metres)

Returning to the earlier BMI example, in determining the body mass index, height and body weight both affect the result. For height, measurements are usually in round centimetres: a recorded measurement of 1.80 metres actually means a height somewhere between 1.795 m and 1.805 m. This uncertainty must be combined with the fluctuation range in weight between 79.5 kg and 80.5 kg. The BMI is defined as the weight in kilograms divided by the square of height in metre. Using either 79.5 kg and 1.795 m or 80.5 kg and 1.805 m gives approximately 24.7. But the person in question may only be 1.795 m tall, with a weight of 80.5 kilograms – or 1.805 m and 79.5 kilograms: all combinations of all possible intermediate values must be considered. Using the interval arithmetic methods described below, the BMI lies in the interval

$[79.5; 80.5]/([1.795; 1.805])^2 = [24.4; 25.0]$.

An operation $\langle \mathrm{op} \rangle$, such as addition or multiplication, on two intervals is defined by

$$[x_1, x_2] \langle \mathrm{op} \rangle [y_1, y_2] = \{x \langle \mathrm{op} \rangle y \mid x \in [x_1, x_2] \,\mathbf{and}\, y \in [y_1, y_2]\}.$$

For the four basic arithmetic operations this can become

$$[x_1, x_2] \langle \mathrm{op} \rangle [y_1, y_2] = [\min(x_1 \langle \mathrm{op} \rangle y_1, x_1 \langle \mathrm{op} \rangle y_2, x_2 \langle \mathrm{op} \rangle y_1, x_2 \langle \mathrm{op} \rangle y_2), \max(x_1 \langle \mathrm{op} \rangle y_1, x_1 \langle \mathrm{op} \rangle y_2, x_2 \langle \mathrm{op} \rangle y_1, x_2 \langle \mathrm{op} \rangle y_2)] \,,$$

provided that $x \langle \mathrm{op} \rangle y$ is allowed for all $x \in [x_1, x_2]$ and $y \in [y_1, y_2]$.

For practical applications this can be simplified further:

- Addition: $[x_1, x_2] + [y_1, y_2] = [x_1 + y_1, x_2 + y_2]$
- Subtraction: $[x_1, x_2] - [y_1, y_2] = [x_1 - y_2, x_2 - y_1]$
- Multiplication: $[x_1, x_2] \cdot [y_1, y_2] = [\min(x_1 y_1, x_1 y_2, x_2 y_1, x_2 y_2), \max(x_1 y_1, x_1 y_2, x_2 y_1, x_2 y_2)]$
- Division: $[x_1, x_2]/[y_1, y_2] = [x_1, x_2] \cdot (1/[y_1, y_2])$, where $1/[y_1, y_2] = [1/y_2, 1/y_1]$ if $0 \notin [y_1, y_2]$

For division by an interval including zero, first define $1/[y_1, 0] = [-\infty, 1/y_1]$ and $1/[0, y_2] = [1/y_2, \infty]$.

For $y_1 < 0 < y_2$, we get $1/[y_1, y_2] = [-\infty, 1/y_1] \cup [1/y_2, \infty]$, which as a single interval gives $1/[y_1, y_2] = [-\infty, \infty]$. This loses useful information about $(1/y_1, 1/y_2)$. So typically it is common to work with $[-\infty, 1/y_1]$ and $[1/y_2, \infty]$ as separate intervals.

Because several such divisions may occur in an interval arithmetic calculation, it is sometimes useful to do the calculation with so-called *multi-intervals* of the form $\bigcup_{i=1}^{l} [x_{i1}, x_{i2}]$. The corresponding *multi-interval arithmetic* maintains a disjoint set of intervals and also provides for overlapping intervals to unite.

Since a real number $r \in \mathbb{R}$ can be interpreted as the interval $[r, r]$, intervals and real numbers can be freely and easily combined.

With the help of these definitions, it is already possible to calculate the range of simple functions, such as $f(a, b, x) = a \cdot x + b$. If, for example, $a = [1, 2]$, $b = [5, 7]$ and $x = [2, 3]$, it is clear

$$f(a, b, x) = ([1, 2] \cdot [2, 3]) + [5, 7] = [1 \cdot 2, 2 \cdot 3] + [5, 7] = [7, 13].$$

Interpreting this as a function $f(a, b, x)$ of the variable x with interval parameters a and b, then it is possible to find the roots of this function. It is then

$$f([1, 2], [5, 7], x) = ([1, 2] \cdot x) + [5, 7] = 0 \Leftrightarrow [1, 2] \cdot x = [-7, -5] \Leftrightarrow x = [-7, -5]/[1, 2]$$

the possible zeros are in the interval $[-7, -2.5]$.

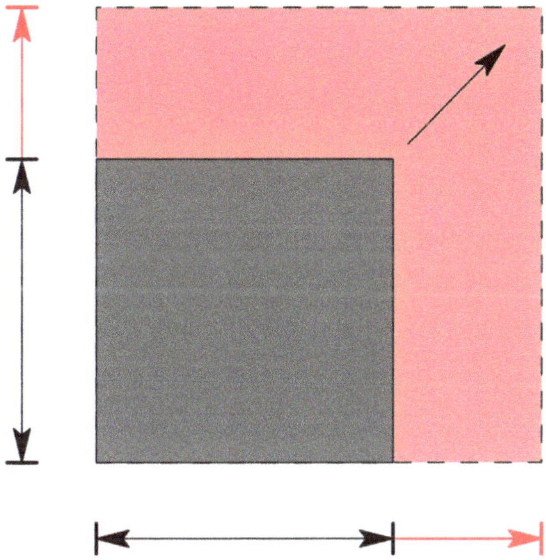

Multiplication of positive intervals

As in the above example, the multiplication of intervals often only requires two multiplications. It is in fact

$[x_1, x_2] \cdot [y_1, y_2] = [x_1 \cdot y_1, x_2 \cdot y_2]$, if $x_1, y_1 \geq 0$.

The multiplication can be seen as a destination area of a rectangle with varying edges. The result interval covers all levels from the smallest to the largest.

The same applies when one of the two intervals is non-positive and the other non-negative. Generally, multiplication can produce results as wide as $[-\infty, \infty]$, for example if $0 \cdot \infty$ is squared. This also occurs, for example, in a division, if the numerator and denominator both contain zero.

Notation

To make the notation of intervals smaller in formulae, brackets can be used.

So we can use $[x] \equiv [x_1, x_2]$ to represent an interval. For the set of all finite intervals, we can use

$[\mathbb{R}] := \{\, [x_1, x_2] \mid x_1 \leq x_2 \text{ and } x_1, x_2 \in \mathbb{R} \cup \{-\infty, \infty\}\}$

as an abbreviation. For a vector of intervals $([x]_1, \ldots, [x]_n) \in [\mathbb{R}]^n$ we can also use a bold font: $[\mathbf{x}]$.

Note that in such a compact notation, $[x]$ should not be confused between a so-called improper or single point interval $[x_1, x_1]$ and the lower and upper limit.

Elementary Functions

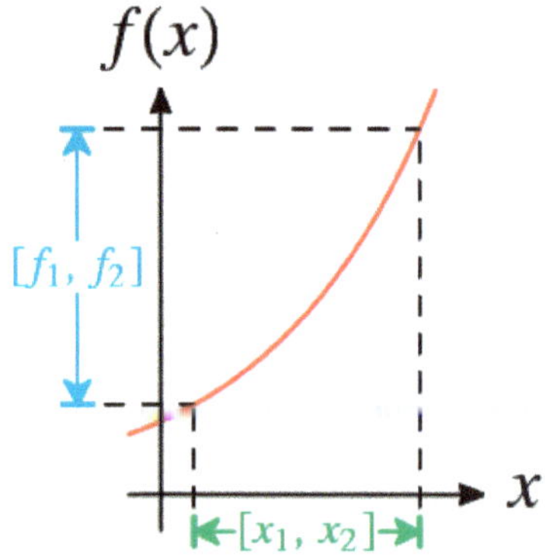

Values of a monotonic function

Interval methods can also apply to functions that do not just use simple arithmetic, and we must also use other basic functions to redefine intervals, using already known monotonicity properties.

For monotonic functions in one variable, the range of values is also easy. If $f : \mathbb{R} \to \mathbb{R}$ is monotonically rising or falling in the interval $[x_1, x_2]$, then for all values in the interval $y_1, y_2 \in [x_1, x_2]$ such that $y_1 \leq y_2$, one of the following inequalities applies:

$f(y_1) \leq f(y_2)$, or $f(y_1) \geq f(y_2)$.

The range corresponding to the interval $[y_1, y_2] \subseteq [x_1, x_2]$ can be calculated by applying the function to the endpoints y_1 and y_2:

$$f([y_1, y_2]) = [\min\{f(y_1), f(y_2)\}, \max\{f(y_1), f(y_2)\}].$$

From this the following basic features for interval functions can easily be defined:

- Exponential function: $a^{[x_1, x_2]} = [a^{x_1}, a^{x_2}]$, for $a > 1$,
- Logarithm: $\log_a([x_1, x_2]) = [\log_a x_1, \log_a x_2]$, for positive intervals $[x_1, x_2]$ and $a > 1$
- Odd powers: $[x_1, x_2]^n = [x_1{}^n, x_2{}^n]$, for odd $n \in \mathbb{N}$.

For even powers, the range of values being considered is important, and needs to be dealt with before doing any multiplication. For example, x^n for $x \in [-1, 1]$ should produce the interval $[0, 1]$ when $n = 2, 4, 6, \ldots$. But if $[-1, 1]^n$ is taken by applying interval multiplication of form x^n for $x \in [-1, 1]$ then the result appears to be $[-1, 1]$, wider than necessary.

Instead consider the function x^n as a monotonically decreasing function for $x < 0$ and a monotonically increasing function for $x > 0$. So for even $n \in \mathbb{N}$:

- $[x_1, x_2]^n = [x_1^n, x_2^n]$, if $x_1 \geq 0$,
- $[x_1, x_2]^n = [x_2^n, x_1^n]$, if $x_2 < 0$,
- $[x_1, x_2]^n = [0, \max\{x_1^n, x_2^n\}]$, otherwise.

More generally, one can say that for piecewise monotonic functions it is sufficient to consider the endpoints x_1, x_2 of the interval $[x_1, x_2]$, together with the so-called *critical points* within the interval being those points where the monotonicity of the function changes direction.

For the sine and cosine functions, the critical points are at $(1/2 + n) \cdot \pi$ or $n \cdot \pi$ for all $n \in \mathbb{Z}$ respectively. Only up to five points matter as the resulting interval is $[-1, 1]$ if the interval includes at least two extrema. For sine and cosine, only the endpoints need full evaluation as the critical points lead to easily pre-calculated values – namely -1, 0, +1.

Interval Extensions of General Functions

In general, it may not be easy to find such a simple description of the output interval for many functions. But it may still be possible to extend functions to interval arithmetic. If $f : \mathbb{R}^n \to \mathbb{R}$ is a function from a real vector to a real number, then $[f] : [\mathbb{R}]^n \to [\mathbb{R}]$ is called an *interval extension* of f if

$$[f]([\mathbf{x}]) \supseteq \{f(\mathbf{y}) | \mathbf{y} \in [\mathbf{x}]\}.$$

This definition of the interval extension does not give a precise result. For example, both {\displaystyle $[f]([x_1, x_2]) = [e^{x_1}, e^{x_2}]$ and $[g]([x_1, x_2]) = [-\infty, \infty]$ are allowable extensions of the exponential function. Extensions as tight as possible are desirable, taking into the relative costs of calculation and imprecision; in this case $[f]$ should be chosen as it give the tightest possible result.

The *natural interval extension* is achieved by combining the function rule $f(x_1, \cdots, x_n)$ with the equivalents of the basic arithmetic and elementary functions.

The *Taylor interval extension* (of degree k) is a $k+1$ is a times differentiable function f defined by

$$[f]([\mathbf{x}]) := f(\mathbf{y}) + \sum_{i=1}^{k} \frac{1}{i!} \mathrm{D}^i f(\mathbf{y}) \cdot ([\mathbf{x}] - \mathbf{y})^i + [r]([\mathbf{x}], [\mathbf{x}], \mathbf{y}),$$

for some $\mathbf{y} \in [\mathbf{x}]$, where $\mathrm{D}^i f(\mathbf{y})$th order differential of f at the point $\mathbf{y}$ and $[r]$ is an interval extension of the *Taylor remainder*

$$r(\mathbf{x}, \xi, \mathbf{y}) = \frac{1}{(k+1)!} \mathrm{D}^{k+1} f(\xi) \cdot (\mathbf{x} - \mathbf{y})^{k+1}.$$

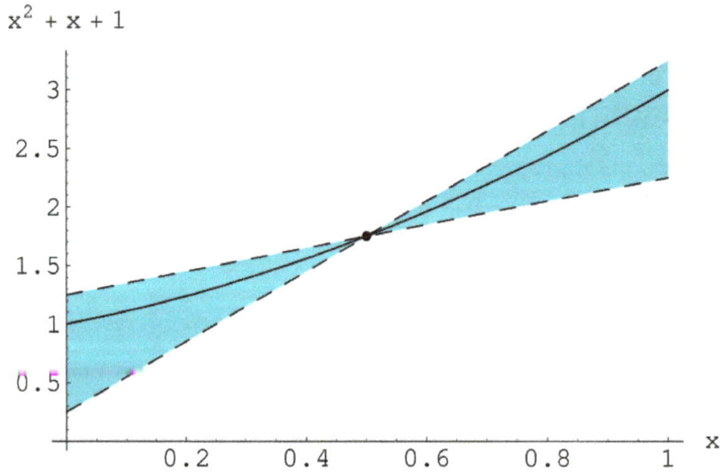

Mean value form

The vector ξ lies between $\mathbf{x}$ and $\mathbf{y}$ with $\mathbf{x}, \mathbf{y} \in [\mathbf{x}]$. ξ is protected by $[\mathbf{x}]$. Usually one chooses $\mathbf{y}$ to be the midpoint of the interval and uses the natural interval extension to assess the remainder.

The special case of the Taylor interval extension of degree $k = 0$ is also referred to as the *mean value form*. For an interval extension of the Jacobian $[J_f]([\mathbf{x}])$ we get

$$[f]([\mathbf{x}]) := f(\mathbf{y}) + [J_f]([\mathbf{x}]) \cdot ([\mathbf{x}] - \mathbf{y}).$$

A nonlinear function can be defined by linear features.

COMPLEX INTERVAL ARITHMETIC

An interval can also be defined as a locus of points at a given distance from the centre, and this definition can be extended from real numbers to complex numbers. As it is the case with computing with real numbers, computing with complex numbers involves uncertain data. So, given the fact that an interval number is a real closed interval and a complex number is an ordered pair of real numbers, there is no reason to limit the application of interval arithmetic to the measure of uncertainties in computations with real numbers. Interval arithmetic can thus be extended, via complex interval numbers, to determine regions of uncertainty in computing with complex numbers.

The basic algebraic operations for real interval numbers (real closed intervals) can be extended to complex numbers. It is therefore not surprising that complex interval arithmetic is similar to, but not the same as, ordinary complex arithmetic. It can be shown that, as it is the case with real interval arithmetic, there is no distributivity between addition and multiplication of complex interval numbers except for certain special cases, and inverse elements do not always exist for complex interval numbers. Two other useful properties of ordinary complex arithmetic fail to hold in complex interval arithmetic: the additive and multiplicative properties, of ordinary complex conjugates, do not hold for complex interval conjugates.

Interval arithmetic can be extended, in an analogous manner, to other multidimensional number systems such as quaternions and octonions, but with the expense that we have to sacrifice other useful properties of ordinary arithmetic.

INTERVAL METHODS

The methods of classical numerical analysis can not be transferred one-to-one into interval-valued algorithms, as dependencies between numerical values are usually not taken into account.

Rounded Interval Arithmetic

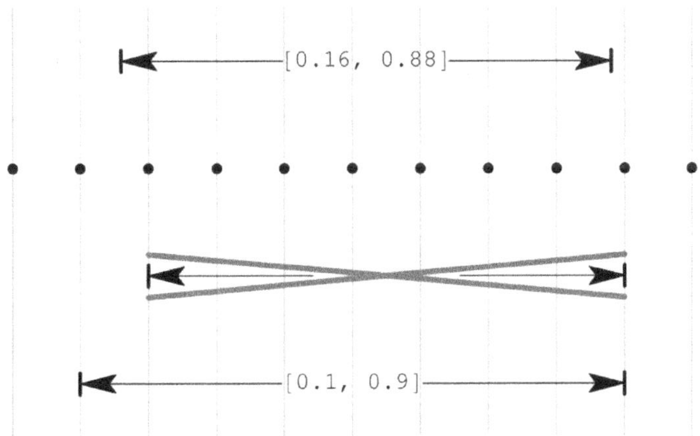

Outer bounds at different level of rounding

To work effectively in a real-life implementation, intervals must be compatible with floating point computing. The earlier operations were based on exact arithmetic, but in general fast numerical solution methods may not be available. The range of values of the function $f(x,y) = x + y$ for $x \in [0.1, 0.8]$ and $y \in [0.06, 0.08]$. Where the same calculation is done with single digit precision, the result would normally be $[0.2, 0.9]$. But $[0.2, 0.9] \not\supseteq [0.16, 0.88]$, so this approach would contradict the basic principles of interval arithmetic, as a part of the domain of $f([0.1, 0.8], [0.06, 0.08])$ would be lost. Instead, the outward rounded solution $[0.1, 0.9]$ is used.

The standard IEEE 754 for binary floating-point arithmetic also sets out procedures for the implementation of rounding. An IEEE 754 compliant system allows programmers to round to the nearest floating point number; alternatives are rounding towards 0 (truncating), rounding toward positive infinity (i.e. up), or rounding towards negative infinity (i.e. down).

The required *external rounding* for interval arithmetic can thus be achieved by changing the rounding settings of the processor in the calculation of the upper limit (up) and lower limit (down). Alternatively, an appropriate small interval $[\varepsilon_1, \varepsilon_2]$ can be added.

Dependency Problem

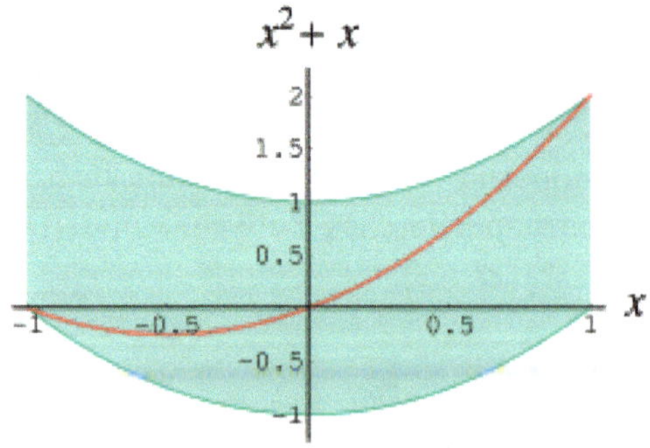

Approximate estimate of the value range

The so-called *dependency problem* is a major obstacle to the application of interval arithmetic. Although interval methods can determine the range of elementary arithmetic operations and functions very accurately, this is not always true with more complicated functions. If an interval occurs several times in a calculation using parameters, and each occurrence is taken independently then this can lead to an unwanted expansion of the resulting intervals.

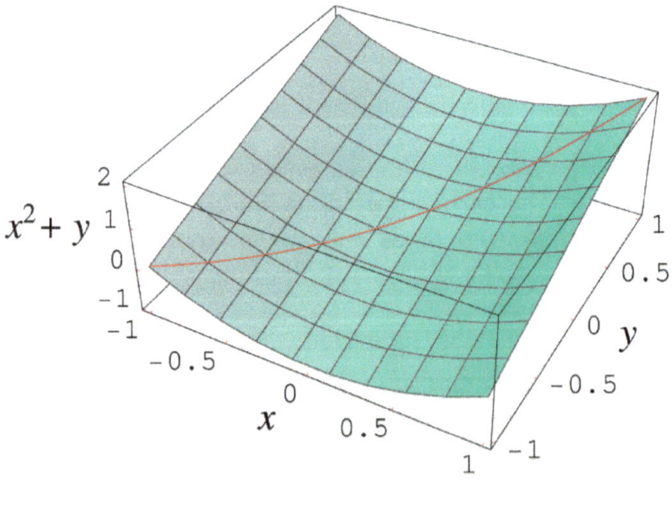

Treating each occurrence of a variable independently

As an illustration, take the function f defined by $f(x) = x^2 + x$. The values of this function over the interval $[-1,1]$ are really $[-1/4, 2]$. As the natural interval extension, it is calculated as $[-1,1]^2 + [-1,1] = [0,1] + [-1,1] = [-1,2]$, which is slightly larger; we have instead calculated the infimum and supremum of the function $h(x,y) = x^2 + y$ over $x, y \in [-1, 1]$. There is a better expression of f in which the variable x only appears once, namely by rewriting $f(x) = x^2 + x$ as addition and squaring in the quadratic {\displaystyle f(x)=\ left(x+{\frac $f(x) = \left(x + \frac{1}{2}\right)^2 - \frac{1}{4}$.

So the suitable interval calculation is

$$\left([-1,1] + \frac{1}{2}\right)^2 - \frac{1}{4} = \left[-\frac{1}{2}, \frac{3}{2}\right]^2 - \frac{1}{4} = \left[0, \frac{9}{4}\right] - \frac{1}{4} = \left[-\frac{1}{4}, 2\right]$$

and gives the correct values.

In general, it can be shown that the exact range of values can be achieved, if each variable appears only once and if f is continuous inside the box. However, not every function can be rewritten this way.

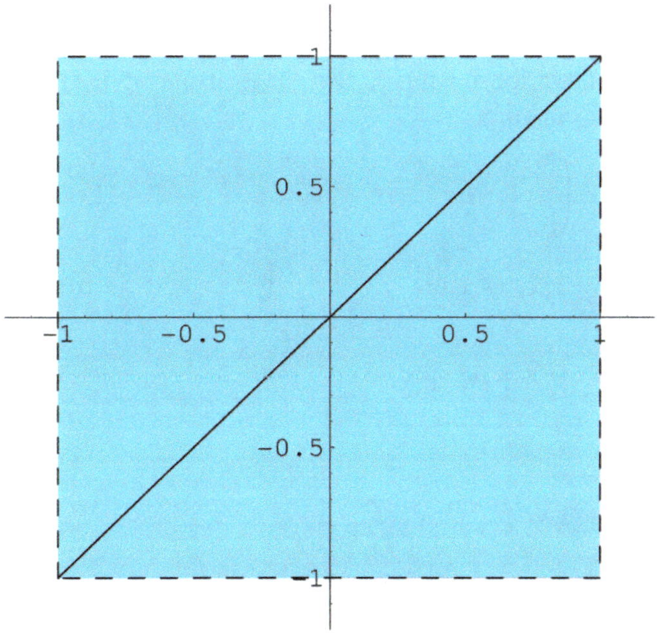

Wrapping effect

The dependency of the problem causing over-estimation of the value range can go as far as covering a large range, preventing more meaningful conclusions.

An additional increase in the range stems from the solution of areas that do not take the form of an interval vector. The solution set of the linear system

$$\begin{aligned} x &= p \\ y &= p \end{aligned}$$

for $p \in [-1, 1]$ is precisely the line between the points $(-1, -1)$ and $(1, 1)$. Using interval methods results in the unit square, $[-1, 1] \times [-1, 1]$. This is known as the *wrapping effect.*

Linear Interval Systems

A linear interval system consists of a matrix interval extension $[\mathbf{A}] \in [\mathbb{R}]^{n\times m}$ and an interval vector $[\mathbf{b}] \in [\mathbb{R}]^{n}$. We want the smallest cuboid $[\mathbf{x}] \in [\mathbb{R}]^{m}$, for all vectors $\mathbf{x} \in \mathbb{R}^{m}$ which there is a pair $(\mathbf{A}, \mathbf{b})$ with $\mathbf{A} \in [\mathbf{A}]$ and $\mathbf{b} \in [\mathbf{b}]$ satisfying

$$\mathbf{A} \cdot \mathbf{x} = \mathbf{b}.$$

For quadratic systems – in other words, for $n = m$ – there can be such an interval vector $[\mathbf{x}]$, which covers all possible solutions, found simply with the interval Gauss method. This replaces the numerical operations, in that the linear algebra method known as Gaussian elimination becomes its interval version. However, since this method uses the interval entities $[\mathbf{A}]$ and $[\mathbf{b}]$ repeatedly in the calculation, it can produce poor results for some problems. Hence using the result of the interval-valued Gauss only provides first rough estimates, since although it contains the entire solution set, it also has a large area outside it.

A rough solution $[\mathbf{x}]$ can often be improved by an interval version of the Gauss–Seidel method. The motivation for this is that the {\displaystyle i}-th row of the interval extension of the linear equation

$$\begin{pmatrix} [a_{11}] & \cdots & [a_{1n}] \\ \vdots & \ddots & \vdots \\ [a_{n1}] & \cdots & [a_{nn}] \end{pmatrix} \cdot \begin{pmatrix} x_1 \\ \vdots \\ x_n \end{pmatrix} = \begin{pmatrix} [b_1] \\ \vdots \\ [b_n] \end{pmatrix}$$

can be determined by the variable x_i if the division $1/[a_{ii}]$ is allowed. It is therefore simultaneously

$$x_j \in [x_j] \text{ and } x_j \in \frac{[b_i] - \sum\limits_{k\neq j} [a_{ik}] \cdot [x_k]}{[a_{ii}]}$$

So we can now replace $[x_j]$ by

$$[x_j] \cap \frac{[b_i] - \sum\limits_{k\neq j} [a_{ik}] \cdot [x_k]}{[a_{ii}]},$$

and so the vector $[\mathbf{x}]$ by each element. Since the procedure is more efficient for a diagonally dominant matrix, instead of the system $[\mathbf{A}] \cdot \mathbf{x} = [\mathbf{b}]$ one can often try multiplying it by an appropriate rational matrix $\mathbf{M}$ with the resulting matrix equation

$$(\mathbf{M} \cdot [\mathbf{A}]) \cdot \mathbf{x} = \mathbf{M} \cdot [\mathbf{b}]$$

left to solve. If one chooses, for example, $\mathbf{M} = \mathbf{A}^{-1}$ for the central matrix $\mathbf{A} \in [\mathbf{A}]$, then $\mathbf{M} \cdot [\mathbf{A}]$, then $\mathbf{A} \in [\mathbf{A}]$, then $\mathbf{M} \cdot [\mathbf{A}]$ is outer extension of the identity matrix.

These methods only work well if the widths of the intervals occurring are sufficiently small. For wider intervals it can be useful to use an interval-linear system on finite (albeit large) real number equivalent linear systems. If all the matrices $\mathbf{A} \in [\mathbf{A}]$ are invertible, it is sufficient to consider all possible combinations (upper and lower) of the endpoints occurring in the intervals. The resulting problems can be resolved using conventional numerical methods. Interval arithmetic is still used to determine rounding errors.

This is only suitable for systems of smaller dimension, since with a fully occupied $n \times n$ matrix, 2^{n^2} real matrices need to be inverted, with 2^n vectors for the right hand side. This approach was developed by Jiri Rohn and is still being developed.

Interval Newton method

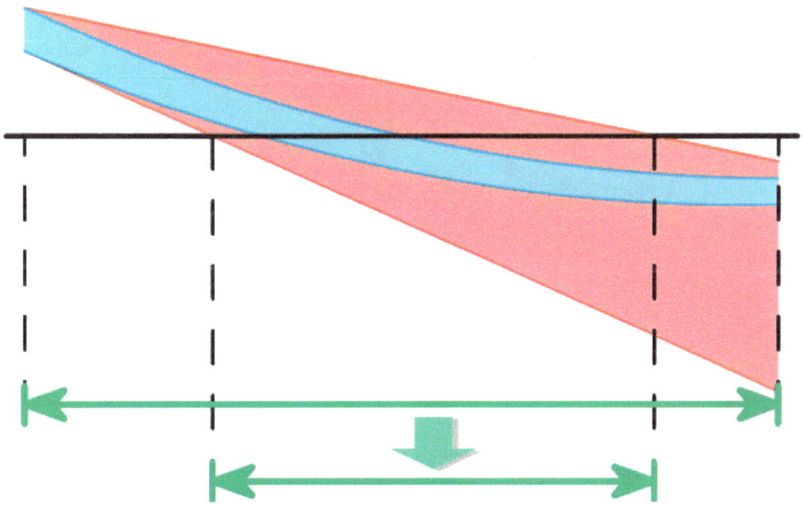

Reduction of the search area in the interval Newton step in "thick" functions

An interval variant of Newton's method for finding the zeros in an interval vector $[\mathbf{x}]$ can be derived from the average value extension. For an unknown vector $\mathbf{z} \in [\mathbf{x}]$ applied to $\mathbf{y} \in [\mathbf{x}]$, gives

$$f(\mathbf{z}) \in f(\mathbf{y}) + [J_f]([\mathbf{x}]) \cdot (\mathbf{z} - \mathbf{y}).$$

For a zero $f(\mathbf{z}) \in f(\mathbf{y}) + [J_f]([\mathbf{x}]) \cdot (\mathbf{z} - \mathbf{y})$. , and thus must satisfy

$$f(\mathbf{y}) + [J_f]([\mathbf{x}]) \cdot (\mathbf{z} - \mathbf{y}) = 0.$$

This is equivalent to $\mathbf{z} \in \mathbf{y} - [J_f]([\mathbf{x}])^{-1} \cdot f(\mathbf{y})$. An outer estimate of $[J_f]([\mathbf{x}])^{-1} \cdot f(\mathbf{y}))$ can be determined using linear methods.

In each step of the interval Newton method, an approximate starting value $[\mathbf{x}] \in [\mathbb{R}]^n$ is replaced by $[\mathbf{x}] \cap (\mathbf{y} - [J_f]([\mathbf{x}])^{-1} \cdot f(\mathbf{y}))$ and so the result can be improved iteratively. In contrast to traditional methods, the interval method approaches the result by containing the zeros. This guarantees that the result produces all zeros in the initial range. Conversely, it proves that no zeros of f were in the initial range $[\mathbf{x}]$ if a Newton step produces the empty set.

The method converges on all zeros in the starting region. Division by zero can lead to separation of distinct zeros, though the separation may not be complete; it can be complemented by the bisection method.

As an example, consider the function $f(x) = x^2 - 2$, the starting range $[x] = [-2, 2]$, and the point $y = 0$. We then have $J_f(x) = 2x$ and the first Newton step gives

$$[-2,2] \cap \left(0 - \frac{1}{2 \cdot [-2,2]}(0-2)\right) = [-2,2] \cap \left([-\infty,-0.5] \cup [0.5,\infty]\right) = [-2,-0.5] \cup [0.5,2].$$

More Newton steps are used separately on $x \in [-2,-0.5]$ and $[0.5, 2]$. These converge to arbitrarily small intervals around $-\sqrt{2}$ and $+\sqrt{2}$.

The Interval Newton method can also be used with *thick functions* such as $g(x) = x^2 - [2,3]$, which would in any case have interval results. The result then produces intervals containing $[-\sqrt{3}, -\sqrt{2}] \cup [\sqrt{2}, \sqrt{3}]$.

Bisection and Covrs

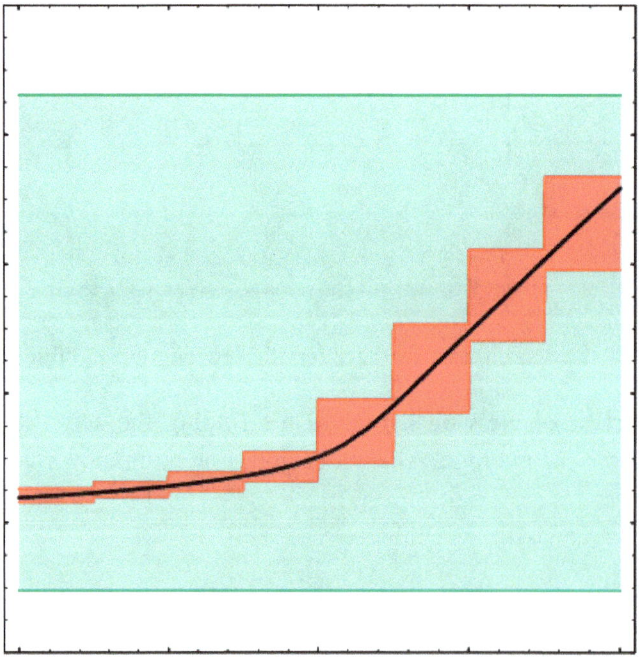

Rough estimate (turquoise) and improved estimates through "mincing" (red)

The various interval methods deliver conservative results as dependencies between the sizes of different intervals extensions are not taken into account. However the dependency problem becomes less significant for narrower intervals.

Covering an interval vector $[\mathbf{x}]$ by smaller boxes $[\mathbf{x}_1], \ldots, [\mathbf{x}_k]$ so that $[\mathbf{x}] = \bigcup_{i=1}^{k} [\mathbf{x}_i]$ is then valid for the range of values $f([\mathbf{x}]) = \bigcup_{i=1}^{k} f([\mathbf{x}_i])$. So for the interval extensions described above, $[f]([\mathbf{x}]) \supseteq \bigcup_{i=1}^{k} [f]([\mathbf{x}_i])$ is valid. Since $[f]([\mathbf{x}])$ is often a genuine superset of the right-hand side, this usually leads to an improved estimate.

Such a cover can be generated by the bisection method such as thick elements $[x_{i1}, x_{i2}]$ of the interval vector $[\mathbf{x}] = ([x_{11}, x_{12}], \ldots, [x_{n1}, x_{n2}])$ by splitting in the centre into the two intervals $[x_{i1}, (x_{i1} + x_{i2})/2]$ and $[(x_{i1} + x_{i2})/2, x_{i2}]$. If the result is still not suitable then further gradual subdivision is possible. Note that a cover of 2^r intervals results from r divisions of vector elements, substantially increasing the computation costs.

With very wide intervals, it can be helpful to split all intervals into several subintervals with a constant (and smaller) width, a method known as *mincing*. This then avoids the calculations for intermediate bisection steps. Both methods are only suitable for problems of low dimension.

APPLICATION

Interval arithmetic can be used in various areas (such as set inversion, motion planning, set estimation or stability analysis) to treat estimates with no exact numerical value.

Rounding Error Analysis

Interval arithmetic is used with error analysis, to control rounding errors arising from each calculation. The advantage of interval arithmetic is that after each operation there is an interval that reliably includes the true result. The distance between the interval boundaries gives the current calculation of rounding errors directly:

Error = $\mathrm{abs}(a - b)$ for a given interval $[a, b]$.

Interval analysis adds to rather than substituting for traditional methods for error reduction, such as pivoting.

Tolerance Analysis

Parameters for which no exact figures can be allocated often arise during the simulation of technical and physical processes. The production process of technical components allows certain tolerances, so some parameters fluctuate within intervals. In addition, many fundamental constants are not known precisely.

If the behavior of such a system affected by tolerances satisfies, for example, $f(\mathbf{x}, \mathbf{p}) = 0$, for $\mathbf{p} \in [\mathbf{p}]$ and unknown x then the set of possible solutions

$$\{\mathbf{x} \mid \exists \mathbf{p} \in [\mathbf{p}], f(\mathbf{x}, \mathbf{p}) = 0\},$$

can be found by interval methods. This provides an alternative to traditional propagation of error analysis. Unlike point methods, such as Monte Carlo simulation, interval arithmetic methodology ensures that no part of the solution area can be overlooked. However, the result is always a worst case analysis for the distribution of error, as other probability-based distributions are not considered.

Fuzzy Interval Arithmetic

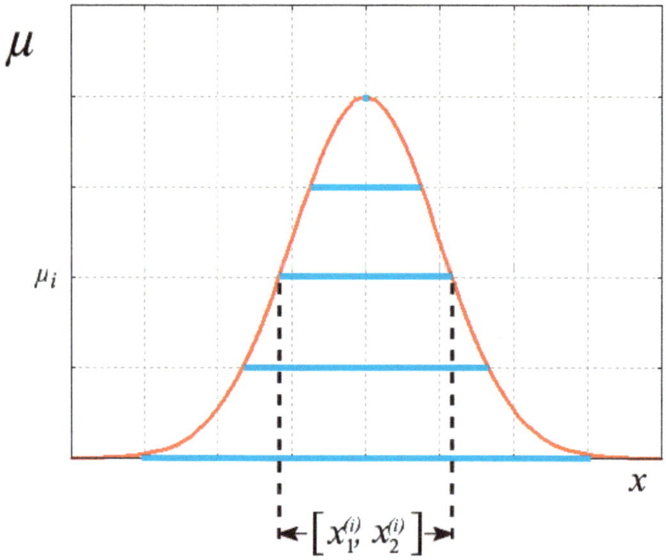

Approximation of the normal distribution by a sequence of intervals

Interval arithmetic can also be used with affiliation functions for fuzzy quantities as they are used in fuzzy logic. Apart from the strict statements $x \in [x]$ and $x \notin [x]$, intermediate values are also possible, to which real numbers $\mu \in [0,1]$ are assigned. $\mu = 1$ corresponds to definite membership while $\mu = 0$ is non-membership. A distribution function assigns uncertainty, which can be understood as a further interval.

For *fuzzy arithmetic* only a finite number of discrete affiliation stages $\mu_i \in [0,1]$ are considered. The form of such a distribution for an indistinct value can then represented by a sequence of intervals

$\left[x^{(1)}\right] \supset \left[x^{(2)}\right] \supset \cdots \supset \left[x^{(k)}\right]$. The interval $[x^{(i)}]$ corresponds exactly to the fluctuation range for the stage μ_i.

The appropriate distribution for a function $f(x_1, \cdots, x_n)$ concerning indistinct values $x_1, \cdots, x_n$ and the corresponding sequences $\left[x_1^{(1)}\right] \supset \cdots \supset \left[x_1^{(k)}\right], \cdots, \left[x_n^{(1)}\right] \supset \cdots \supset \left[x_n^{(k)}\right]$ can be approximated by the sequence $\left[y^{(1)}\right] \supset \cdots \supset \left[y^{(k)}\right]$. The values $\left[y^{(i)}\right]$ are given by $\left[y^{(i)}\right] = f\left(\left[x_1^{(i)}\right], \cdots \left[x_n^{(i)}\right]\right)$ and can be calculated by interval methods. The value $\left[y^{(1)}\right]$ corresponds to the result of an interval calculation.

HISTORY

Interval arithmetic is not a completely new phenomenon in mathematics; it has appeared several times under different names in the course of history. For example, Archimedes calculated lower and upper bounds $223/71 < \pi < 22/7$ in the 3rd century BC. Actual calculation with intervals has neither been as popular as other numerical techniques, nor been completely forgotten.

Rules for calculating with intervals and other subsets of the real numbers were published in a 1931 work by Rosalind Cicely Young, a doctoral candidate at the University of Cambridge. Arithmetic work on range numbers to improve reliability of digital systems were then published in a 1951 textbook on linear algebra by Paul Dwyer (University of Michigan); intervals were used to measure rounding errors associated with floating-point numbers. A comprehensive paper on interval algebra in numerical analysis was published by Teruo Sunaga (1958).

The birth of modern interval arithmetic was marked by the appearance of the book *Interval Analysis* by Ramon E. Moore in 1966. He had the idea in Spring 1958, and a year later he published an article about computer interval arithmetic. Its merit was that starting with a simple principle, it provided a general method for automated error analysis, not just errors resulting from rounding.

Independently in 1956, Mieczyslaw Warmus suggested formulae for calculations with intervals,though Moore found the first non-trivial applications.

In the following twenty years, German groups of researchers carried out pioneering work around Götz Alefeld (de) and Ulrich Kulisch (de) at the University of Karlsruhe and later also at the Bergische University of Wuppertal. For example, Karl Nickel (de) explored more effective implementations, while improved containment procedures for the solution set of systems of equations were due to Arnold Neumaier among others. In the 1960s, Eldon R. Hansen dealt with interval extensions for linear equations and then provided crucial contributions to global optimisation, including what is now known as Hansen"s method, perhaps the most widely used interval algorithm. Classical methods in this often have the problem of determining the largest (or smallest) global value, but could only find a local optimum and could not find better values; Helmut Ratschek and Jon George Rokne developed branch and bound methods, which until then had only applied to integer values, by using intervals to provide applications for continuous values.

In 1988, Rudolf Lohner developed Fortran-based software for reliable solutions for initial value problems using ordinary differential equations.

The journal *Reliable Computing* (originally *Interval Computations*) has been published since the 1990s, dedicated to the reliability of computer-aided computations. As lead editor, R. Baker Kearfott, in addition to his work on global optimisation, has contributed significantly to the unification of notation and terminology used in interval arithmetic (Web: Kearfott). In recent years work has concentrated in particular on the estimation of preimages of parameterised functions and to robust control theory by the COPRIN working group of INRIA in Sophia Antipolis in France (Web: INRIA).

IMPLEMENTATIONS

There are many software packages that permit the development of numerical applications using interval arithmetic. These are usually provided in the form of program libraries. There are also C++ and Fortran compilers that handle interval data types and suitable operations as a language extension, so interval arithmetic is supported directly.

Since 1967, *Extensions for Scientific Computation* (XSC) have been developed in the University of Karlsruhe for various programming languages, such as C++, Fortran and Pascal. The first platform was a Zuse Z 23, for which a new interval data type with appropriate elementary operators was made available. There followed in 1976, Pascal-SC, a Pascal variant on a Zilog Z80 that it made possible to create fast, complicated routines for automated result verification. Then came the Fortran 77-based ACRITH-XSC for the System/370 architecture (FORTRAN-SC), which was later delivered by IBM. Starting from 1991 one could produce code for C compilers with Pascal-XSC; a year later the C++ class library supported C-XSC on many different computer systems. In 1997, all XSC variants were made available under the GNU General Public License. At the beginning of 2000 C-XSC 2.0 was released under the leadership of the working group for scientific computation at the Bergische University of Wuppertal to correspond to the improved C++ standard.

Another C++-class library was created in 1993 at the Hamburg University of Technology called *Profil/BIAS* (Programmer's Runtime Optimized Fast Interval Library, Basic Interval Arithmetic), which made the usual interval operations more user friendly. It emphasized the efficient use of hardware, portability and independence of a particular presentation of intervals.

The Boost collection of C++ libraries contains a template class for intervals. Its authors are aiming to have interval arithmetic in the standard C++ language.

Gaol is another C++ interval arithmetic library that is unique in that it offers the relational interval operators used in interval constraint programming.

The Frink programming language has an implementation of interval arithmetic that handles arbitrary-precision numbers. Programs written in Frink can use intervals without rewriting or recompilation.

In addition computer algebra systems, such as Mathematica, Maple and MuPAD, can handle intervals. A Matlab extension *Intlab* builds on BLAS routines, and the Toolbox b4m makes a Profil/BIAS interface. Moreover, the Software Euler Math Toolbox includes an interval arithmetic.

A library for the functional language OCaml was written in assembly language and C.

IEEE Std 1788-2015 – IEEE standard for interval arithmetic

A standard for interval arithmetic has been approved in June 2015. Two reference implementations are freely available. These have been developed by members of the standard"s working group: The libieeep1788 library for C++, and the interval package for GNU Octave.

A minimal subset of the standard is currently under development that should be easier to implement and may speed production of implementations.

CONFERENCES AND WORKSHOP

Several international conferences or workshop take place every year in the world. The main conference is probably SCAN (International Symposium on Scientific Computing,

Computer Arithmetic, and Verified Numerical Computation), but there is also SWIM (Small Workshop on Interval Methods), PPAM (International Conference on Parallel Processing and Applied Mathematics), REC (International Workshop on Reliable Engineering Computing).

REFERENCES

1. Kulisch, Ulrich (1989). Wissenschaftliches Rechnen mit Ergebnisverifikation. Eine Einführung *(in German). Wiesbaden:* Vieweg-Verlag. ISBN 3-528-08943-1.
2. Dreyer, Alexander (2003). Interval Analysis of Analog Circuits with Component Tolerances. Aachen, Germany: Shaker Verlag. ISBN 3-8322-4555-3.
3. Complex interval arithmetic and its applications, Miodrag Petkovi?, Ljiljana Petkovi?, Wiley-VCH, 1998, ISBN 978-3-527-40134-5
4. Hend Dawood (2011). Theories of Interval Arithmetic: Mathematical Foundations and Applications. Saarbrücken: LAP LAMBERT Academic Publishing. ISBN 978-3-8465-0154-2.
5. Jiri Rohn, List of publications
6. Walster, G. William; Hansen, Eldon Robert (2004). Global Optimization using Interval Analysis (2nd ed.). New York, USA: Marcel Dekker. ISBN 0-8247-4059-9.
7. Jaulin, Luc; Kieffer, Michel; Didrit, Olivier; Walter, Eric (2001). Applied Interval Analysis. Berlin: Springer. ISBN 1-85233-219-0.
8. Application of Fuzzy Arithmetic to Quantifying the Effects of Uncertain Model Parameters, Michael Hanss, University of Stuttgart
9. Sunaga, Teruo (1958). Theory of interval algebra and its application to numerical analysis. RAAG Memoirs. pp. 29–46.
10. Moore, R. E. (1966). Interval Analysis. Englewood Cliff, New Jersey, USA: Prentice-Hall. ISBN 0-13-476853-1.
11. Cloud, Michael J.; Moore, Ramon E.; Kearfott, R. Baker (2009). Introduction to Interval Analysis. Philadelphia: Society for Industrial and Applied Mathematics (SIAM). ISBN 0-89871-669-1.
12. Hansen, E. R. (2001-08-13). "Publications Related to Early Interval Work of R. E. Moore". University of Louisiana at Lafayette Press. Retrieved 2015-06-29.
13. Precursory papers on interval analysis by M. Warmus Archived April 18, 2008, at the Wayback Machine.
14. Alefeld, Götz; Herzberger, Jürgen. Einführung in die Intervallrechnung. Reihe Informatik (in German). **12**. Mannheim, Wien, Zürich: B.I.-Wissenschaftsverlag. ISBN 3-411-01466-0.
15. Kulisch, Ulrich (1969). "Grundzüge der Intervallrechnung". In Laugwitz, Detlef. Jahrbuch *Überblicke* Mathematik *(in German).* **2**. Mannheim, Germany: Bibliographisches Institut. pp. 51–98.
16. Bounds for ordinary differential equations of Rudolf Lohner (in German)
17. Software for Interval Computations collected by Vladik Kreinovich , University of Texas at El Paso

18. History of XSC-Languages
19. A Proposal to add Interval Arithmetic to the C++ Standard Library
20. Gaol is Not Just Another Interval Arithmetic Library
21. INTerval LABoratory and b4m
22. Alliot, Jean-Marc; Gotteland, Jean-Baptiste; Vanaret, Charlie; Durand, Nicolas; Gianazza, David (2012), Implementing an interval computation library for OCaml on x86/amd64 architectures, 17th ACM SIGPLAN International Conference on Functional Programming
23. IEEE Standard for Interval Arithmetic
24. Revol, Nathalie (2015). The (near-)future IEEE 1788 standard for interval arithmetic. 8th small workshop on interval methods. Slides (PDF)
25. C++ implementation of the preliminary IEEE P1788 standard for interval arithmetic
26. GNU Octave interval package
27. P1788.1 - Standard for Interval Arithmetic (Simplified)". IEEE Project. IEEE Standards Association. Retrieved 2016-08-16.

CHAPTER 7

Fuzzy Measure Theory

Chapter Contents

- **Introduction**
- **Definitions**
- **Simplification Assumptions For Fuzzy Measures**

Sugeno (1974) introduced the theory of fuzzy measures and fuzzy integrals. A fuzzy measure g over a set X (the universe of discource with the subsets E, F, ...) satifics the following conditions (X is finite):

1) $g(\varnothing) = 0, \quad g(X) = 1$

2) If $E \subset F$, then $g(E) \leq g(F)$

A fuzzy measure is a Sugeno measure (or a -fuzzy measure) if it satifies the following additional condition for some > -1:

$$g_\lambda(E \cup F) = g_\lambda(E) + g_\lambda(F) + \lambda \cdot g_\lambda(E) \cdot g_\lambda(F)$$

The value of can be calculated regarding to the condition g(X)=1:

$$\lambda + 1 = \prod_{i=1}^{n} \left(1 + \lambda g^i\right).$$

INTRODUCTION

In mathematics, fuzzy measure theory considers generalized measures in which the additive property is replaced by the weaker property of monotonicity. The central concept of fuzzy measure theory is the fuzzy measure (also *capacity*, see) which was introduced by Choquet in 1953 and independently defined by Sugeno in 1974 in the context of fuzzy integrals. There exists a number of different classes of fuzzy measures including plausibility/belief measures; possibility/ necessity measures; and probability measures which are a subset of classical measures.

DEFINITIONS

Let Xbe a universe of discourse, C be a class of subsets of $\mathbf{X}$, and $E, F \in \mathcal{C}$. A function $g: \mathcal{C} \to \mathbb{R}$ where

1. $\emptyset \in \mathcal{C} \Rightarrow g(\emptyset) = 0$
2. $E \subseteq F \Rightarrow g(E) \leq g(F)$

is called a *fuzzy measure*. A fuzzy measure is called *normalized* or *regular* if $g(\mathbf{X}) = 1$.

Properties of fuzzy measures

For any $E, F \in \mathcal{C}$, a fuzzy measure is:

- **additive** if $g(E \cup F) = g(E) + g(F)$. for all $E \cap F = \emptyset$;
- **supermodular** if $g(E \cup F) + g(E \cap F) \geq g(E) + g(F)$;
- **submodular** if $g(E \cup F) + g(E \cap F) \leq g(E) + g(F)$;
- **superadditive** if $g(E \cup F) \geq g(E) + g(F)$ for all $E \cap F = \emptyset$;
- **subadditive** if $g(E \cup F) \leq g(E) + g(F)$ for all $E \cap F = \emptyset$;
- **symmetric** if $|E| = |F|$ implies $g(E) = g(F)$;
- **Boolean** if $g(E) = 0$ or $g(E) = 1$.

Understanding the properties of fuzzy measures is useful in application. When a fuzzy measure is used to define a function such as the Sugeno integral or Choquet integral, these properties will be crucial in understanding the function"s behavior. For instance, the Choquet integral with respect to an additive fuzzy measure reduces to the Lebesgue integral. In discrete cases, a symmetric fuzzy measure will result in the ordered weighted averaging (OWA) operator. Submodular fuzzy measures result in convex functions, while supermodular fuzzy measures result in concave functions when used to define a Choquet integral.

Möbius representation

Let *g* be a fuzzy measure, the Möbius representation of *g* is given by the set function *M*, where for every $E, F \subseteq X$,

$$M(E) = \sum_{F \subseteq E} (-1)^{|E \setminus F|} g(F).$$

The equivalent axioms in Möbius representation are:

1. $M(\emptyset) = 0$.
2. $\sum_{F \subseteq E | i \in F} M(F) \geq 0$, for all $E \subseteq \mathbf{X}$ and all $i \in E$

A fuzzy measure in Möbius representation M is called *normalized* if $\sum_{E \subseteq \mathbf{X}} M(E) = 1.$

Möbius representation can be used to give an indication of which subsets of $\mathbf{X}$ interact with one another. For instance, an additive fuzzy measure has Möbius values all equal to zero except for singletons. The fuzzy measure g in standard representation can be recovered from the Möbius form using the Zeta transform:

$$g(E) = \sum_{F \subseteq E} M(F), \forall E \subseteq \mathbf{X}.$$

SIMPLIFICATION ASSUMPTIONS FOR FUZZY MEASURES

Fuzzy measures are defined on a semiring of sets or monotone class which may be as granular as the power set of **X**, and even in discrete cases the number of variables can as large as $2^{|\mathbf{X}|}$. For this reason, in the context of multi-criteria decision analysis and other disciplines, simplification assumptions on the fuzzy measure have been introduced so that it is less computationally expensive to determine and use. For instance, when it is assumed the fuzzy measure is *additive*, it will hold that $g(E) = \sum_{i \in E} g(\{i\})$ and the values of the fuzzy measure can be evaluated from the values on **X**. Similarly, a *symmetric* fuzzy measure is defined uniquely by |**X**| values. Two important fuzzy measures that can be used are the Sugeno- or λ-fuzzy measure and *k*-additive measures, introduced by Sugeno and Grabischrespectively.

Sugeno λ-Measure

The Sugeno λ-measure is a special case of fuzzy measures defined iteratively. It has the following definition:

Definition

Let $\mathbf{X} = \{x_1, \ldots, x_n\}$ be a finite set and let $\lambda \in (-1, +\infty)$. A **Sugeno λ-measure** is a function [0,1] such that

1. $g(X) = 1$.
2. if $A, B \subseteq \mathbf{X}$ (alternatively $A, B \in 2^{\mathbf{X}}$) with $A \cap B = \emptyset$ then $g(A \cup B) = g(A) + g(B) + \lambda g(A) g(B)$.

As a convention, the value of g at a singleton set $\{x_i\}$ is called a density and is denoted by $g_i = g(\{x_i\})$. In addition, we have that λ satisfies the property

$$\lambda + 1 = \prod_{i=1}^{n} (1 + \lambda g_i).$$

Tahani and Keller as well as Wang and Klir have showed that once the densities are known, it is possible to use the previous polynomial to obtain the values of λ uniquely.

k-additive Fuzzy Measure

The *k*-additive fuzzy measure limits the interaction between the subsets X to size $E \subseteq X$. This drastically reduces the number of variables needed to define the fuzzy measure, and as *k* can be anything from 1 (in which case the fuzzy measure is additive) to **X**, it allows for a compromise between modelling ability and simplicity.

A discrete fuzzy measure *g* on a set **X** is called *k-additive* $(1 \leq k \leq |\mathbf{X}|)$ if its Möbius representation verifies $M(E) = 0$, whenever $|E| > k$ for any $E \subseteq \mathbf{X}$, and there exists a subset *F* with *k* elements such that $M(F) \neq 0$.

Shapley and Interaction Indices

In game theory, the **Shapley value**, named in honour of Lloyd Shapley, who introduced it in 1953, is a solution concept in cooperative game theory. To each cooperative game it assigns a unique distribution (among the players) of a total surplus generated by the coalition of all players. The Shapley value is characterized by a collection of desirable properties. Hart (1989) provides a survey of the subject.

The setup is as follows: a coalition of players cooperates, and obtains a certain overall gain from that cooperation. Since some players may contribute more to the coalition than others or may possess different bargaining power (for example threatening to destroy the whole surplus), what final distribution of generated surplus among the players should arise in any particular game? Or phrased differently: how important is each player to the overall cooperation, and what payoff can he or she reasonably expect? The Shapley value provides one possible answer to this question.

Formal Definition

Formally, a **coalitional game** is defined as: There is a set *N* (of *n* players) and a function v that maps subsets of players to the real numbers: $v : 2^N \to \mathbb{R}$, with $v(\emptyset) = 0$, where $\emptyset$ denotes the empty set. The function $v(\emptyset) = 0$, where $\emptyset$ is called a characteristic function.

The function v has the following meaning: if *S* is a coalition of players, then $v(S)$, called the worth of coalition *S*, describes the total expected sum of payoffs the members of s can obtain by cooperation.

The Shapley value is one way to distribute the total gains to the players, assuming that they all collaborate. It is a "fair" distribution in the sense that it is the only distribution with certain desirable properties listed below. According to the Shapley value, the amount that player *i* gets given in a coalitional game (v, N) is

$$\phi_i(v) = \sum_{S \subseteq N \setminus \{i\}} \frac{|S|!\,(n - |S| - 1)!}{n!}(v(S \cup \{i\}) - v(S))$$

where *n* is the total number of players and the sum extends over all subsets *S* of *N* not containing player *i*. The formula can be interpreted as follows: imagine the coalition being formed one actor at a time, with each actor demanding their contribution $v(S \cup \{i\}) - v(S)$

as a fair compensation, and then for each actor take the average of this contribution over the possible different permutations in which the coalition can be formed.

An alternative equivalent formula for the Shapley value is:

$$\phi_i(v) = \frac{1}{|N|!} \sum_R \left[v(P_i^R \cup \{i\}) - v(P_i^R)\right]$$

where the sum ranges over all $|N|!$ orders R of the players and P_i^R is the set of players in N which precede i in the order R.

Example

Consider a simplified description of a business. An owner, o, provides crucial capital in the sense that without him no gains can be obtained. There are k workers $w_1,\ldots,w_k$, each of whom contributes an amount p to the total profit. So $N = \{o, w_1,\ldots,w_k\}$ and $v(S) = 0$ if o is not a member of S and $v(S) = mp$ if S contains the owner and m workers. Computing the Shapley value for this coalition game leads to a value of kp/2 for the owner and p/2 for each worker.

Glove Game

The glove game is a coalitional game where the players have left and right hand gloves and the goal is to form pairs.

$$N = \{1, 2, 3\}$$

where players 1 and 2 have right hand gloves and player 3 has a left hand glove

The value function for this coalitional game is

$$v(S) = \begin{cases} 1, & \text{if } S \in \{\{1,3\}, \{2,3\}, \{1,2,3\}\} \\ 0, & \text{otherwise} \end{cases}$$

Where the formula for calculating the Shapley value is:

$$\phi_i(v) = \frac{1}{|N|!} \sum_R \left[v(P_i^R \cup \{i\}) - v(P_i^R)\right]$$

Where R is an ordering of the players and P_i^R is the set of players in N which precede i in the order R

The following table displays the marginal contributions of Player 1

Order R	MC_1
$1,2,3$	$v(\{1\}) - v(\varnothing) = 0 - 0 = 0$
$1,3,2$	$v(\{1\}) - v(\varnothing) = 0 - 0 = 0$
$2,1,3$	$v(\{1,2\}) - v(\{2\}) = 0 - 0 = 0$
$2,3,1$	$v(\{1,2,3\}) - v(\{2,3\}) = 1 - 1 = 0$
$3,1,2$	$v(\{1,3\}) - v(\{3\}) = 1 - 0 = 1$
$3,2,1$	$v(\{1,2,3\}) - v(\{2,3\}) = 1 - 1 = 0$

$$\phi_1(v) = \left(\frac{1}{6}\right)(1) = \frac{1}{6}$$

By a symmetry argument it can be shown that

$$\phi_2(v) = \phi_1(v) = \frac{1}{6}$$

Due to the efficiency axiom the sum of all the Shapley values is equal to 1, which means that

$$\phi_3(v) = \frac{4}{6} = \frac{2}{3}.$$

Properties

The Shapley value has the following desirable properties:

1. Efficiency: The total gain is distributed:

$$\sum_{i \in N} \phi_i(v) = v(N)$$

2. *Symmetry*: If *i* and *j* are two actors who are equivalent in the sense that

$$v(S \cup \{i\}) = v(S \cup \{j\})$$

for every subset *S* of *N* which contains neither *i* nor *j*, then $\phi_i(v) = \phi_j(v)$.

3. *Linearity*: If two coalition games described by gain functions *v* and *w* are combined, then the distributed gains should correspond to the gains derived from *v* and the gains derived from *w*:

$$\phi_i(v + w) = \phi_i(v) + \phi_i(w)$$

for every *i* in *N*. Also, for any real number *a*,

$$\phi_i(av) = a\phi_i(v)$$

for every *i* in *N*.

4. *Zero player (null player)*: The Shapley value $\phi_i(v)$ of a null player i in a game v is zero. A player $v(S \cup \{i\}) = v(S)$ for all coalitions S.

Given a player set *N*, the Shapley value is the only map from the set of all games to payoff vectors that satisfies *all four* properties 1, 2, 3, and 4 from above.

Addendum Definitions

1. ***Anonymous:*** If *i* and *j* are two actors, and *w* is the gain function that acts just like *v* except that the roles of *i* and *j* have been exchanged, then $\phi_i(v) = \phi_j(w)$. In essence, this means that the labeling of the actors doesn't play a role in the assignment of their gains. Such a function is said to be *anonymous*.

2. *Marginalism*: the Shapley value can be defined as a function which uses only the marginal contributions of player i as the arguments.

Aumann–Shapley Value

In their 1974 book, Lloyd Shapley and Robert Aumann extended the concept of the Shapley value to infinite games (defined with respect to a non-atomic measure), creating the diagonal formula. This was later extended by Jean-François Mertens and Abraham Neyman.

As seen above, the value of an n-person game associates to each player the expectation of his contribution to the worth or the coalition or players before him in a random ordering of all the players. When there are many players and each individual plays only a minor role, the set of all players preceding a given one is heuristically thought as a good sample of the players so that the value of a given infinitesimal player {\displaystyle ds} around as "his" contribution to the worth of a "perfect" sample of the population of all players.

Symbolically, if v is the coalitional worth function associating to each coalition c measured subset of a measurable set I that can be thought as $I = [0, 1]$ without loss of generality.

$$(Sv)(ds) = \int_0^1 (v(tI + ds) - v(tI))dt.$$

where $(Sv)(ds)$denotes the Shapley value of the infinitesimal player ds in the game, tI is a perfect sample of the all-player set I containing a proportion t of all the players, and $tI + ds$ is the coalition obtained after ds joins tI. This is the heuristic

This is the heuristic form of the diagonal formula.

Assuming some regularity of the worth function, for example assuming v can be represented as differentiable function of a non-atomic measure on $I, \mu, v(c) = f(\mu(c))$ with density function ϕ, with $\mu(c) = \int 1_c(u)\phi(u)du$, ($1_c()$ the characteristic function of c. Under such conditions. Under such conditions

$$\mu(tI) = t\mu(I),$$

as can be shown by approximating the density by a step function and keeping the proportion t for each level of the density function, and

$$v(tI + ds) = f(t\mu(I)) + f'(t\mu(I))\mu(ds).$$

The diagonal formula has then the form developed by Aumann and Shapley (1974)

$$(Sv)(ds) = \int_0^1 f'_{t\mu(I)}(\mu(ds))dt$$

Above μ can be vector valued (as long as the function is defined and differentiable on the range of μ, the above formula makes sense).

In the argument above if the measure contains atoms $\mu(tI) = t\mu(I)$ is no longer true—this is why the diagonal formula mostly applies to non-atomic games.

Two approaches were deployed to extend this diagonal formula when the function f is no longer differentiable. Mertens goes back to the original formula and takes the derivative after the integral thereby benefiting from the smoothing effect. Neyman took a different approach. Going back to an elementary application of Mertens"s approach from Mertens (1980):

$$(Sv)(ds) = \lim_{\epsilon \to 0, \epsilon > 0} \frac{1}{\epsilon} \int_0^{1-\epsilon} (f(t + \epsilon\mu(ds)) - f(t))dt$$

This works for example for majority games—while the original diagonal formula cannot be used directly. How Mertens further extends this by identifying symmetries that the Shapley value should be invariant upon, and averaging over such symmetries to create further smoothing effect commuting averages with the derivative operation as above. A survey for non atomic value is found in Neyman (2002)

REFERENCES

1. Shapley, Lloyd S. (1953). "A Value for n-person Games". In Kuhn, H. W.; Tucker, A. W. Contributions to the Theory of Games. Annals of Mathematical Studies. 28. Princeton University Press. pp. 307–317.
2. Roth, Alvin E., ed. (1988). The Shapley Value: Essays in Honor of Lloyd S. Shapley. Cambridge: Cambridge University Press. ISBN 0-521-36177-X.
3. Hart, Sergiu (1989). "Shapley Value". In Eatwell, J.; Milgate, M.; Newman, P. The New Palgrave: Game Theory. Norton. pp. 210–216.
4. Hart, Sergiu (May 12, 2016). "A Bibliography of Cooperative Games: Value Theory".
5. For a proof of unique existence, see Ichiishi, Tatsuro (1983). Game Theory for Economic Analysis. New York: Academic Press. pp. 118–120. ISBN 0-12-370180-5.
6. Aumann, Robert J.; Shapley, Lloyd S. (1974). Values of Non-Atomic Games. Princeton: Princeton Univ. Press. ISBN 0-691-08103-4.
7. Mertens, Jean-François (1980). "Values and Derivatives". Mathematics of Operations Research. 5 (4): 523–552. doi:10.1287/moor.5.4.523. JSTOR 3689325.
8. Mertens, Jean-François (1988). "The Shapley Value in the Non Differentiable Case". International Journal of Game Theory. 17 (1): 1–65. doi:10.1007/BF01240834.
9. Neyman, A., 2002. Value of Games with infinitely many Players, "Handbook of Game Theory with Economic Applications," Handbook of Game Theory with Economic Applications, Elsevier, edition 1, volume 3, number 3, 00. R.J. Aumann & S. Hart (ed.).

CHAPTER 8

Field-Programmable Gate Array

Chapter Contents

- **Introducton**
- **Technical Design**
- **History**
- **Comparisons**
- **Applications**
- **Architectur**
- **Design And Programming**
- **Basic Process Technology Types**
- **Major Manufacturers**
- **Hardware Description Language**
- **Static Random-Access Memory**
- **Flash Memory**

A field-programmable gate array (FPGA) is an integrated circuit (IC) that can be programmed in the field after manufacture. FPGAs are similar in principle to, but have vastly wider potential application than, programmable read-only memory (PROM) chips. FPGAs are used by engineers in the design of specialized ICs that can later be produced hard-wired in large quantities for distribution to computer manufacturers and end users. Ultimately, FPGAs might allow computer users to tailor microprocessors to meet their own individual needs.

INTRODUCTON

A field-programmable gate array (FPGA) is an integrated circuit designed to be configured by a customer or a designer after manufacturing – hence "field-programmable". The FPGA configuration is generally specified using a hardware description language (HDL), similar to that used for an application-specific integrated circuit (ASIC). (Circuit diagrams were previously used to specify the configuration, as they were for ASICs, but this is increasingly rare.)

A Stratix IV FPGA from Altera

A Spartan FPGA from Xilinx

FPGAs contain an array of programmable logic blocks, and a hierarchy of reconfigurable interconnects that allow the blocks to be "wired together", like many logic gates that can be inter-wired in different configurations. Logic blocks can be configured to perform complex combinational functions, or merely simple logic gates like AND and XOR. In most FPGAs, logic blocks also include memory elements, which may be simple flip-flops or more complete blocks of memory.

TECHNICAL DESIGN

Contemporary field-programmable gate arrays (FPGAs) have large resources of logic gates and RAM blocks to implement complex digital computations. As FPGA designs employ very fast I/Os and bidirectional data buses, it becomes a challenge to verify correct timing of valid data within setup time and hold time. Floor planning enables resources allocation within FPGAs to meet these time constraints. FPGAs can be used to implement any logical function that an ASIC could perform. The ability to update the functionality after shipping, partial re-configuration of a portion of the design and the low non-recurring engineering costs relative to an ASIC design (notwithstanding the generally higher unit cost), offer advantages for many applications.

Some FPGAs have analog features in addition to digital functions. The most common analog feature is programmable slew rate on each output pin, allowing the engineer to set low rates on lightly loaded pins that would otherwise ring or couple unacceptably, and to set higher rates on heavily loaded pins on high-speed channels that would otherwise run too slowly.Also common are quartz-crystal oscillators, on-chip resistance-capacitance oscillators, and phase-locked loops with embedded voltage-controlled oscillators used for clock generation and management and for high-speed serializer-deserializer (SERDES) transmit clocks and receiver clock recovery. Fairly common are differential comparators on input pins designed to be connected to differential signaling channels. A few "mixed signal FPGAs" have integrated peripheral analog-to-digital converters (ADCs) and digital-to-analog converters (DACs) with analog signal conditioning blocks allowing them to operate as a system-on-a-chip. Such devices blur the line between an FPGA, which carries digital ones and zeros on its internal programmable interconnect fabric, and field-programmable analog array (FPAA), which carries analog values on its internal programmable interconnect fabric.

HISTORY

The FPGA industry sprouted from programmable read-only memory (PROM) and programmable logic devices (PLDs). PROMs and PLDs both had the option of being programmed in batches in a factory or in the field (field-programmable). However, programmable logic was hard-wired between logic gates.

In the late 1980s, the Naval Surface Warfare Center funded an experiment proposed by Steve Casselman to develop a computer that would implement 600,000 reprogrammable gates. Casselman was successful and a patent related to the system was issued in 1992.

Some of the industry's foundational concepts and technologies for programmable logic arrays, gates, and logic blocks are founded in patents awarded to David W. Page and LuVerne R. Peterson in 1985.

Altera was founded in 1983 and delivered the industry's first reprogrammable logic device in 1984 – the EP300 – which featured a quartz window in the package that allowed users to shine an ultra-violet lamp on the die to erase the EPROM cells that held the device configuration. Xilinx co-founders Ross Freeman and Bernard Vonderschmitt invented the first commercially viable field-programmable gate array in 1985 – the XC2064. The XC2064 had programmable gates and programmable interconnects between gates, the beginnings of a new technology and market. The XC2064 had 64 configurable logic blocks (CLBs), with two three-input lookup tables (LUTs). More than 20 years later, Freeman was entered into the National Inventors Hall of Fame for his invention.

Altera and Xilinx continued unchallenged and quickly grew from 1985 to the mid-1990s, when competitors sprouted up, eroding significant market share. By 1993, Actel (now Microsemi) was serving about 18 percent of the market. By 2010, Altera (31 percent), Actel (10 percent) and Xilinx (36 percent) together represented approximately 77 percent of the FPGA market.

The 1990s were an explosive period of time for FPGAs, both in sophistication and the volume of production. In the early 1990s, FPGAs were primarily used in telecommunications and networking. By the end of the decade, FPGAs found their way into consumer, automotive, and industrial applications.

Modern Developments

A recent trend has been to take the coarse-grained architectural approach a step further by combining the logic blocks and interconnects of traditional FPGAs with embedded microprocessors and related peripherals to form a complete "system on a programmable chip". This work mirrors the architecture by Ron Perlof and Hana Potash of Burroughs Advanced Systems Group which combined a reconfigurable CPU architecture on a single chip called the SB24. That work was done in 1982. Examples of such hybrid technologies can be found in the Xilinx Zynq-7000 All Programmable SoC, which includes a 1.0 GHz dual-core ARM Cortex-A9 MPCore processor embedded within the FPGA's logic fabric or in the Altera Arria V FPGA, which includes an 800 MHz dual-core ARM Cortex-A9 MPCore. The Atmel FPSLIC is another such device, which uses an AVR processor in combination with Atmel's programmable logic architecture. The Microsemi SmartFusion devices incorporate an ARM Cortex-M3 hard processor core (with up to 512 kB of flash and 64 kB of RAM) and analog peripherals such as a multi-channel ADC and DACs to their flash-based FPGA fabric.

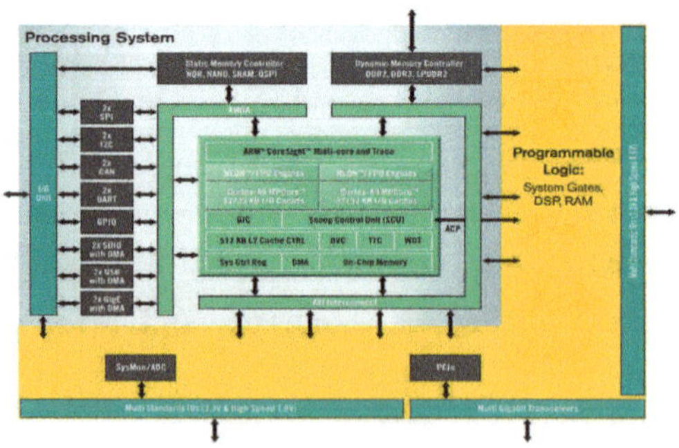

A Xilinx Zynq-7000 All Programmable System on a Chip.

In 2010, Xilinx Inc introduced the first All Programmable System on a Chip branded Zynq™-7000 that fused features of an ARM high-end microcontroller (hard-core implementations of a 32-bit processor, memory, and I/O) with a 28 nm FPGA fabric to make it easier for embedded designers to use. The extensible processing platform enables system architects and embedded software developers to apply a combination of serial and parallel processing to their embedded system designs, for which the general trend has been to progressively increasing complexity. The high level of integration helps to reduce power consumption and dissipation, and the reduced parts count versus using an FPGA with a separate CPU chip leads to a lower parts cost, a smaller system, and higher reliability since most failures in modern electronics occur on PCBs in the connections between chips instead of within the chips themselves.

An alternate approach to using hard-macro processors is to make use of soft processor cores that are implemented within the FPGA logic. Nios II, MicroBlaze and Mico32 are examples of popular softcore processors. Many modern FPGAs are programmed at "run time", and this is leading to the idea of reconfigurable computing or reconfigurable systems – CPUs that reconfigure themselves to suit the task at hand. Additionally, new, non-FPGA architectures are beginning to emerge. Software-configurable microprocessors such as the Stretch S5000 adopt a hybrid approach by providing an array of processor cores and FPGA-like programmable cores on the same chip.

Companies like Microsoft have started to use FPGA to accelerate high-performance, computationally intensive systems (like the data centers that operate their Bing search engine), due to the performance per Watt advantage FPGAs deliver.

Gates

- 1982: 8192 gates, Burroughs Advances Systems Group, integrated into the S-Type 24-bit processor for reprogrammable I/O.
- 1987: 9,000 gates, Xilinx
- 1992: 600,000, Naval Surface Warfare Department
- Early 2000s: Millions

Market size

- 1985: First commercial FPGA : Xilinx XC2064
- 1987: $14 million
- ≈1993: >$385 million
- 2005: $1.9 billion
- 2010 estimates: $2.75 billion
- 2013: $5.4 billion
- 2020 estimate: $9.8 billion

Design starts

- 2005: 80,000
- 2008: 90,000

COMPARISONS

Historically, FPGAs have been slower, less energy efficient and generally achieved less functionality than their fixed ASIC counterparts. An older study had shown that designs implemented on FPGAs need on average 40 times as much area, draw 12 times as much dynamic power, and run at one third the speed of corresponding ASIC implementations. More recently, FPGAs such as the Xilinx Virtex-7 or the Altera Stratix 5 have come to rival corresponding ASIC and ASSP solutions by providing significantly reduced power, increased speed, lower materials cost, minimal implementation real-estate, and increased possibilities for re-configuration 'on-the-fly'. Where previously a design may have included 6 to 10 ASICs, the same design can now be achieved using only one FPGA.

Advantages of FPGAs include the ability to re-program in the field to fix bugs, and may include a shorter time to market and lower non-recurring engineering costs. Vendors can also take a middle road by developing their hardware on ordinary FPGAs, but manufacture their final version as an ASIC so that it can no longer be modified after the design has been committed.

Xilinx claims that several market and technology dynamics are changing the ASIC/FPGA paradigm:

- Integrated circuit development costs are rising aggressively
- ASIC complexity has lengthened development time
- R&D resources and headcount are decreasing
- Revenue losses for slow time-to-market are increasing
- Financial constraints in a poor economy are driving low-cost technologies

These trends make FPGAs a better alternative than ASICs for a larger number of higher-volume applications than they have been historically used for, to which the company attributes the growing number of FPGA design starts (see History).

Some FPGAs have the capability of partial re-configuration that lets one portion of the device be re-programmed while other portions continue running.

Complex Programmable Logic Devices (CPLD)

The primary differences between CPLDs (complex programmable logic devices) and FPGAs are architectural. A CPLD has a somewhat restrictive structure consisting of one or more programmable sum-of-products logic arrays feeding a relatively small number of clocked registers. The result of this is less flexibility, with the advantage of more predictable timing delays and a higher logic-to-interconnect ratio. The FPGA architectures, on the other hand, are dominated by interconnect. This makes them far more flexible (in terms of the range of designs that are practical for implementation within them) but also far more complex to design for.

In practice, the distinction between FPGAs and CPLDs is often one of size as FPGAs are usually much larger in terms of resources than CPLDs. Typically only FPGAs contain more complex embedded functions such as adders, multipliers, memory, and serdes. Another common distinction is that CPLDs contain embedded flash to store their configuration while FPGAs usually, but not always, require external nonvolatile memory.

Security Considerations

With respect to security, FPGAs have both advantages and disadvantages as compared to ASICs or secure microprocessors. FPGAs' flexibility makes malicious modifications during fabrication a lower risk. Previously, for many FPGAs, the design bitstream was exposed while the FPGA loads it from external memory (typically on every power-on). All major FPGA vendors now offer a spectrum of security solutions to designers such as bitstream encryption and authentication. For example, Altera and Xilinx offer AES (up to 256 bit) encryption for bitstreams stored in an external flash memory.

FPGAs that store their configuration internally in nonvolatile flash memory, such as Microsemi's ProAsic 3 or Lattice's XP2 programmable devices, do not expose the bitstream and do not need encryption. In addition, flash memory for a lookup table provides single event upset protection for space applications. Customers wanting a higher guarantee of tamper resistance can use write-once, Antifuse FPGAs from vendors such as Microsemi.

With its Stratix 10 FPGAs and SoCs, Altera introduced a Secure Device Manager and physically uncloneable functions to provide high levels of protection against physical attacks.

APPLICATIONS

An FPGA can be used to solve any problem which is computable. This is trivially proven by the fact FPGA can be used to implement a soft microprocessor, such as the Xilinx MicroBlaze or Altera Nios II. Their advantage lies in that they are sometimes significantly faster for some applications because of their parallel nature and optimality in terms of the number of gates used for a certain process.

Specific applications of FPGAs include digital signal processing, software-defined radio, ASIC prototyping, medical imaging, computer vision, speech recognition, cryptography, bioinformatics, computer hardware emulation, radio astronomy, metal detection and a growing range of other areas.

FPGAs originally began as competitors to CPLDs and competed in a similar space, that of glue logic for PCBs. As their size, capabilities, and speed increased, they began to take over larger and larger functions to the point where some are now marketed as full systems on chips (SoC). Particularly with the introduction of dedicated multipliers into FPGA architectures in the late 1990s, applications which had traditionally been the sole reserve of DSPs began to incorporate FPGAs instead.

Another trend on the usage of FPGAs is hardware acceleration, where one can use the FPGA to accelerate certain parts of an algorithm and share part of the computation between the FPGA and a generic processor.

Traditionally, FPGAs have been reserved for specific vertical applications where the volume of production is small. For these low-volume applications, the premium that companies pay in hardware costs per unit for a programmable chip is more affordable than the development resources spent on creating an ASIC for a low-volume application. Today, new cost and performance dynamics have broadened the range of viable applications.

Common FPGA Applications:

- Aerospace and Defense
    - Avionics/DO-254
    - Communications
    - Missiles & Munitions
    - Secure Solutions
    - Space
- Medical Electronics
- ASIC Prototyping
- Audio
    - Connectivity Solutions
    - Portable Electronics
    - Radio
    - Digital Signal Processing (DSP)
- Automotive
    - High Resolution Video
    - Image Processing
    - Vehicle Networking and Connectivity
    - Automotive Infotainment
- Broadcast
    - Real-Time Video Engine
    - EdgeQAM
    - Encoders
    - Displays
    - Switches and Routers
- Consumer Electronics
    - Digital Displays
    - Digital Cameras
    - Multi-function Printers
    - Portable Electronics
    - Set-top Boxes
- Data Center
    - Servers
    - Security

    - HSM
    - Routers
    - Switches
    - Gatcways
    - Load Balancing
- High Performance Computing
    - Servers
    - Super Computers
    - SIGINT Systems
    - High-end RADARs
    - High-end Beam Forming Systems
    - Data Mining Systems
- Industrial
    - Industrial Imaging
    - Industrial Networking
    - Motor Control
- Medical
    - Ultrasound
    - CT Scanner
    - MRI
    - X-ray
    - PET
    - Surgical Systems
- Scientific Instruments
    - Lock-in amplifiers
    - Boxcar averagers
    - Phase-locked loops
- Security
    - Industrial Imaging
    - Secure Solutions
    - HSM
    - Image Processing
- Video & Image Processing
    - High Resolution Video
    - Video Over IP Gateway
    - Digital Displays

  - Industrial Imaging
- Wired Communications
  - Optical Transport Networks
  - Network Processing
  - Connectivity Interfaces
- Wireless Communications
  - Baseband
  - Connectivity Interfaces
  - Mobile Backhaul
  - Radio

ARCHITECTURE

Logic Blocks

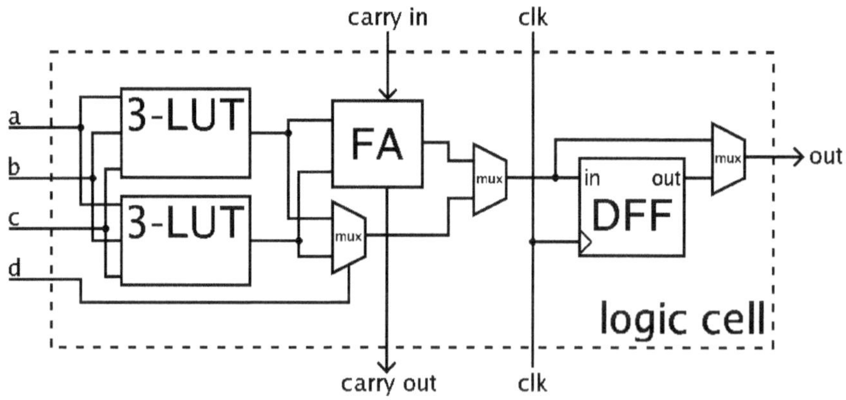

Simplified example illustration of a logic cell

The most common FPGA architecture consists of an array of logic blocks (called configurable logic block, CLB, or logic array block, LAB, depending on vendor), I/O pads, and routing channels. Generally, all the routing channels have the same width (number of wires). Multiple I/O pads may fit into the height of one row or the width of one column in the array.

An application circuit must be mapped into an FPGA with adequate resources. While the number of CLBs/LABs and I/Os required is easily determined from the design, the number of routing tracks needed may vary considerably even among designs with the same amount of logic. For example, a crossbar switch requires much more routing than a systolic array with the same gate count. Since unused routing tracks increase the cost (and decrease the performance) of the part without providing any benefit, FPGA manufacturers try to provide just enough tracks so that most designs that will fit in terms of lookup tables (LUTs) and I/Os can be routed. This is determined by estimates such as those derived from Rent's rule or by experiments with existing designs.

In general, a logic block (CLB or LAB) consists of a few logical cells (called ALM, LE, slice etc.). A typical cell consists of a 4-input LUT, a full adder (FA) and a D-type flip-flop, as shown below. The LUTs are in this figure split into two 3-input LUTs. In *normal mode* those are combined into a 4-input LUT through the left mux. In *arithmetic* mode, their outputs are fed to the FA. The selection of mode is programmed into the middle multiplexer. The output can be either synchronous or asynchronous, depending on the programming of the mux to the right, in the figure example. In practice, entire or parts of the FA are put as functions into the LUTs in order to save space.

Hard blocks

Modern FPGA families expand upon the above capabilities to include higher level functionality fixed into the silicon. Having these common functions embedded into the silicon reduces the area required and gives those functions increased speed compared to building them from primitives. Examples of these include multipliers, generic DSP blocks, embedded processors, high speed I/O logic and embedded memories.

Higher-end FPGAs can contain high speed multi-gigabit transceivers and *hard IP cores* such as processor cores, Ethernet MACs, PCI/PCI Express controllers, and external memory controllers. These cores exist alongside the programmable fabric, but they are built out of transistors instead of LUTs so they have ASIC level performance and power consumption while not consuming a significant amount of fabric resources, leaving more of the fabric free for the application-specific logic. The multi-gigabit transceivers also contain high performance analog input and output circuitry along with high-speed serializers and deserializers, components which cannot be built out of LUTs. Higher-level PHY layer functionality such as line coding may or may not be implemented alongside the serializers and deserializers in hard logic, depending on the FPGA.

Clocking

Most of the circuitry built inside of an FPGA is synchronous circuitry that requires a clock signal. FPGAs contain dedicated global and regional routing networks for clock and reset so they can be delivered with minimal skew. Also, FPGAs generally contain analog PLL and/or DLL components to synthesize new clock frequencies as well as attenuate jitter. Complex designs can use multiple clocks with different frequency and phase relationships, each forming separate clock domains. These clock signals can be generated locally by an oscillator or they can be recovered from a high speed serial data stream. Care must be taken when building clock domain crossing circuitry to avoid metastability. FPGAs generally contain block RAMs that are capable of working as dual port RAMs with different clocks, aiding in the construction of building FIFOs and dual port buffers that connect differing clock domains.

3D Architectures

To shrink the size and power consumption of FPGAs, vendors such as Tabula and Xilinx have introduced new 3D or stacked architectures. Following the introduction of its 28 nm 7-series FPGAs, Xilinx revealed that several of the highest-density parts in those FPGA product lines will be constructed using multiple dies in one package, employing technology developed for 3D construction and stacked-die assemblies.

Xilinx's approach stacks several (three or four) active FPGA die side-by-side on a silicon interposer – a single piece of silicon that carries passive interconnect. The multi-die construction also allows different parts of the FPGA to be created with different process technologies, as the process requirements are different between the FPGA fabric itself and the very high speed 28 Gbit/s serial transceivers. An FPGA built in this way is called a *heterogeneous FPGA*.

Altera's heterogeneous approach involves using a single monolithic FPGA die and connecting other die/technologies to the FPGA using Intel's embedded multi-die interconnect bridge (EMIB) technology.

DESIGN AND PROGRAMMING

To define the behavior of the FPGA, the user provides a hardware description language (HDL) or a schematic design. The HDL form is more suited to work with large structures because it's possible to just specify them numerically rather than having to draw every piece by hand. However, schematic entry can allow for easier visualisation of a design.

Then, using an electronic design automation tool, a technology-mapped netlist is generated. The netlist can then be fit to the actual FPGA architecture using a process called place-and-route, usually performed by the FPGA company's proprietary place-and-route software. The user will validate the map, place and route results via timing analysis, simulation, and other verification methodologies. Once the design and validation process is complete, the binary file generated (also using the FPGA company's proprietary software) is used to (re)configure the FPGA. This file is transferred to the FPGA/CPLD via a serial interface (JTAG) or to an external memory device like an EEPROM.

The most common HDLs are VHDL and Verilog, although in an attempt to reduce the complexity of designing in HDLs, which have been compared to the equivalent of assembly languages, there are moves to raise the abstraction level through the introduction of alternative languages. National Instruments' LabVIEW graphical programming language (sometimes referred to as "G") has an FPGA add-in module available to target and program FPGA hardware.

To simplify the design of complex systems in FPGAs, there exist libraries of predefined complex functions and circuits that have been tested and optimized to speed up the design process. These predefined circuits are commonly called *IP cores*, and are available from FPGA vendors and third-party IP suppliers (rarely free, and typically released under proprietary licenses). Other predefined circuits are available from developer communities such as OpenCores (typically released under free and open source licenses such as the GPL, BSD or similar license), and other sources.

In a typical design flow, an FPGA application developer will simulate the design at multiple stages throughout the design process. Initially the RTL description in VHDL or Verilog is simulated by creating test benches to simulate the system and observe results. Then, after the synthesis engine has mapped the design to a netlist, the netlist is translated to a gate level description where simulation is repeated to confirm the synthesis proceeded without errors. Finally the design is laid out in the FPGA at which point propagation delays can be added and the simulation run again with these values back-annotated onto the netlist.

More recently, OpenCL is being used by programmers to take advantage of the performance and power efficiencies that FPGAs provide. OpenCL allows programmers to develop code in the C programming language and target FPGA functions as OpenCL kernels using OpenCL constructs.

BASIC PROCESS TECHNOLOGY TYPES

- SRAM – based on static memory technology. In-system programmable and re-programmable. Requires external boot devices. CMOS. Currently in use. It is worth noting that flash or EEPROM devices may often load contents into internal SRAM that controls routing and logic.
- Fuse – One-time programmable. Bipolar. Obsolete.
- Antifuse – One-time programmable. CMOS.
- PROM – Programmable Read-Only Memory technology. One-time programmable because of plastic packaging. Obsolete.
- EPROM – Erasable Programmable Read-Only Memory technology. One-time programmable but with window, can be erased with ultraviolet (UV) light. CMOS. Obsolete.
- EEPROM – Electrically Erasable Programmable Read-Only Memory technology. Can be erased, even in plastic packages. Some but not all EEPROM devices can be in-system programmed. CMOS.
- Flash – Flash-erase EPROM technology. Can be erased, even in plastic packages. Some but not all flash devices can be in-system programmed. Usually, a flash cell is smaller than an equivalent EEPROM cell and is therefore less expensive to manufacture. CMOS.

MAJOR MANUFACTURERS

Xilinx and Altera are the current FPGA market leaders and long-time industry rivals. Together, they control over 80 percent of the market. Both Xilinx and Altera provide proprietaryWindows and Linux design software (ISE/Vivado and Quartus) which enable engineers to design, analyse, simulate and synthesize (compile) their designs.

Other manufacturers include:

- Lattice Semiconductor (SRAM based with integrated configuration flash, instant-on, low power, live reconfiguration)

- Microsemi (previously Actel, antifuse, flash-based, mixed-signal)
- SiliconBlue Technologies (extremely low power SRAM-based FPGAs with optional integrated nonvolatile configuration memory; acquired by Lattice in 2011)
- Achronix (SRAM based, 1.5 GHz fabric speed),
- QuickLogic (Ultra Low Power Sensor Hubs, extremely low powered, low density SRAM-based FPGAs, Display bridges MIPI & RGB inputs, MIPI, RGB and LVDS outputs)
- e2v (previously QP semiconductor)
- Atmel (second source of some Altera-compatible devices; also FPSLIC mentioned above)

In March 2010, Tabula announced their FPGA technology that uses time-multiplexed logic and interconnect that claims potential cost savings for high-density applications. On March 24, 2015, Tabula officially shut down.

On June 1, 2015, Intel announced it would acquire Altera for approximately $16.7 billion and completed the acquisition on December 30, 2015.

HARDWARE DESCRIPTION LANGUAGE

In electronics, a **hardware description language (HDL)** is a specialized computer language used to describe the structure and behavior of electronic circuits, and most commonly, digital logic circuits.

A hardware description language enables a precise, formal description of an electronic circuit that allows for the automated analysis and simulation of an electronic circuit. It also allows for the synthesis of a HDL description into a netlist (a specification of physical electronic components and how they are connected together), which can then be placed and routed to produce the set of masks used to create an integrated circuit.

A hardware description language looks much like a programming language such as C; it is a textual description consisting of expressions, statements and control structures. One important difference between most programming languages and HDLs is that HDLs explicitly include the notion of time.

HDLs form an integral part of electronic design automation (EDA) systems, especially for complex circuits, such as application-specific integrated circuits, microprocessors, and programmable logic devices.

Motivation

Due to the exploding complexity of digital electronic circuits since the 1970s (see Moore's law), circuit designers needed digital logic descriptions to be performed at a high level without being tied to a specific electronic technology, such as CMOS or BJT. HDLs were created to implement register-transfer level abstraction, a model of the data flow and timing of a circuit. There are two major hardware description languages: VHDL and Verilog. There are different types of description in them "dataflow, behavioral and structural". Example of dataflow of hdl:

```
library iEEE;
use iEEE.STD_LOGIC_1164.ALL;
use iEEE.STD_NUMERIC_STD.ALL;
entity notl is
  port(a:in STD_LOGIC;
       b:out STD_logic);
end notl;
architecture behavioral of notl is
begin
 b <= not a;
end behavioral;
```

Structure of HDL

HDLs are standard text-based expressions of the structure of electronic systems and their behaviour over time. Like concurrent programming languages, HDL syntax and semantics include explicit notations for expressing concurrency. However, in contrast to most software programming languages, HDLs also include an explicit notion of time, which is a primary attribute of hardware. Languages whose only characteristic is to express circuit connectivity between a hierarchy of blocks are properly classified as netlist languages used in electric computer-aided design (CAD). HDL can be used to express designs in structural, behavioral or register-transfer-level architectures for the same circuit functionality; in the latter two cases the synthesizer decides the architecture and logic gate layout.

HDLs are used to write executable specifications for hardware. A program designed to implement the underlying semantics of the language statements and simulate the progress of time provides the hardware designer with the ability to model a piece of hardware before it is created physically. It is this executability that gives HDLs the illusion of being programming languages, when they are more precisely classified as specification languages or modeling languages. Simulators capable of supporting discrete-event (digital) and continuous-time (analog) modeling exist, and HDLs targeted for each are available.

Comparison with Control-flow Languages

It is certainly possible to represent hardware semantics using traditional programming languages such as C++, which operate on control flow semantics as opposed to data flow, although to function as such, programs must be augmented with extensive and unwieldy class libraries. Generally, however, software programming languages do not include any capability for explicitly expressing time, and thus cannot function as hardware description languages. Before the introduction of System Verilog in 2002, C++ integration with a logic simulator was one of the few ways to use object-oriented programming in hardware verification. System Verilog is the first major HDL to offer object orientation and garbage collection. Using the proper subset of hardware description language, a program called a synthesizer, or logic synthesis tool, can infer hardware logic operations from the language statements and produce an equivalent netlist of generic hardware primitives to implement the specified behaviour. Synthesizers generally ignore the expression of any timing constructs in the text. Digital logic synthesizers, for example, generally use clock edges as the way to time the circuit, ignoring any timing constructs. The ability to have a synthesizable subset of the language does not itself make a hardware description language.

The first hardware description languages appeared in the late 1960s, looking like more traditional languages. The first that had a lasting effect was described in 1971 in C. Gordon Bell and Allen Newell"s text *Computer Structures*. This text introduced the concept of register transfer level, first used in the ISP language to describe the behavior of the Digital Equipment Corporation (DEC) PDP-8.

The language became more widespread with the introduction of DEC"s PDP-16 RT-Level Modules (RTMs) and a book describing their use. At least two implementations of the basic ISP language (ISPL and ISPS) followed. ISPS was well suited to describe relations between the inputs and the outputs of the design and was quickly adopted by commercial teams at DEC, as well as by a number of research teams both in the USA and among its NATO allies.

The RTM products never took off commercially and DEC stopped marketing them in the mid-1980s, as new techniques and in particular very-large-scale integration (VLSI) became more popular.

Separate work done about 1979 at the University of Kaiserslautern produced a language called KARL ("KAiserslautern Register Transfer Language"), which included design calculus language features supporting VLSI chip floorplanning and structured hardware design. This work was also the basis of KARL's interactive graphic sister language ABL, whose name was an initialism for "A Block diagram Language". ABL was implemented in the early 1980s by the Centro Studi e Laboratori Telecomunicazioni (CSELT) in Torino, Italy, producing the ABLED graphic VLSI design editor. In the mid-1980s, a VLSI design framework was implemented around KARL and ABL by an international consortium funded by the Commission of the European Union.

By the late 1970s, design using programmable logic devices (PLDs) became popular, although these designs were primarily limited to designing finite state machines. The work at Data General in 1980 used these same devices to design the Data General Eclipse MV/8000, and commercial need began to grow for a language that could map well to them. By 1983 Data I/O introduced ABEL to fill that need.

As design shifted to VLSI, the first modern HDL, Verilog, was introduced by Gateway Design Automation in 1985. Cadence Design Systems later acquired the rights to Verilog-XL, the HDL simulator that would become the de facto standard of Verilog simulators for the next decade. In 1987, a request from the U.S. Department of Defense led to the development of VHDL (VHSIC Hardware Description Language). VHDL was based on the Ada programming language, as well as on the experience gained with the earlier development of ISPS. Initially, Verilog and VHDL were used to document and simulate circuit designs already captured and described in another form (such as schematic files). HDL simulation enabled engineers to work at a higher level of abstraction than simulation at the schematic level, and thus increased design capacity from hundreds of transistors to thousands.

The introduction of logic synthesis for HDLs pushed HDLs from the background into the foreground of digital design. Synthesis tools compiled HDL source files (written in a constrained format called RTL) into a manufacturable netlist description in terms of gates and transistors. Writing synthesizable RTL files required practice and discipline on the part of the designer; compared to a traditional schematic layout, synthesized RTL netlists were almost always larger in area and slower in performance. A circuit design from a skilled engineer, using labor-intensive schematic-capture/hand-layout, would almost always outperform its logically-synthesized equivalent, but the productivity advantage held by synthesis soon displaced digital schematic capture to exactly those areas that were problematic for RTL synthesis: extremely high-speed, low-power, or asynchronous circuitry.

Within a few years, VHDL and Verilog emerged as the dominant HDLs in the electronics industry, while older and less capable HDLs gradually disappeared from use. However, VHDL and Verilog share many of the same limitations: neither is suitable for analog or mixed-signal circuit simulation; neither possesses language constructs to describe recursively-generated logic structures. Specialized HDLs (such as Confluence) were introduced with the explicit goal of fixing specific limitations of Verilog and VHDL, though none were ever intended to replace them.

Over the years, much effort has been invested in improving HDLs. The latest iteration of Verilog, formally known as IEEE 1800-2005 SystemVerilog, introduces many new features (classes, random variables, and properties/assertions) to address the growing need for better test bench randomization, design hierarchy, and reuse. A future revision of VHDL is also in development, and is expected to match SystemVerilog's improvements.

Design using HDL

As a result of the efficiency gains realized using HDL, a majority of modern digital circuit design revolves around it. Most designs begin as a set of requirements or a high-level architectural diagram. Control and decision structures are often prototyped in flowchart applications, or entered in a state diagram editor. The process of writing the HDL description is highly dependent on the nature of the circuit and the designer's preference for coding style. The HDL is merely the 'capture language', often beginning with a high-level algorithmic description such as a C++ mathematical model. Designers often use scripting languages such as Perl to automatically generate repetitive circuit structures in the HDL language. Special text editors offer features for automatic indentation, syntax-dependent coloration, and macro-based expansion of the entity/architecture/signal declaration.

The HDL code then undergoes a code review, or auditing. In preparation for synthesis, the HDL description is subject to an array of automated checkers. The checkers report deviations from standardized code guidelines, identify potential ambiguous code constructs before they can cause misinterpretation, and check for common logical coding errors, such as floating ports or shorted outputs. This process aids in resolving errors before the code is synthesized.

In industry parlance, HDL design generally ends at the synthesis stage. Once the synthesis tool has mapped the HDL description into a gate netlist, the netlist is passed off to the back-end stage. Depending on the physical technology (FPGA, ASIC gate array, ASIC standard cell), HDLs may or may not play a significant role in the back-end flow. In general, as the design flow progresses toward a physically realizable form, the design database becomes progressively more laden with technology-specific information, which cannot be stored in a generic HDL description. Finally, an integrated circuit is manufactured or programmed for use.

Simulating and Debugging HDL code

Essential to HDL design is the ability to simulate HDL programs. Simulation allows an HDL description of a design (called a model) to pass design verification, an important milestone that validates the design's intended function (specification) against the code implementation in the HDL description. It also permits architectural exploration. The engineer can experiment with design choices by writing multiple variations of a base design, then comparing their behavior in simulation. Thus, simulation is critical for successful HDL design.

To simulate an HDL model, an engineer writes a top-level simulation environment (called a test bench). At minimum, a testbench contains an instantiation of the model (called the device under test or DUT), pin/signal declarations for the model's I/O, and a clock waveform. The testbench code is event driven: the engineer writes HDL statements to implement the (testbench-generated) reset-signal, to model interface transactions (such as a host–bus read/write), and to monitor the DUT's output. An HDL simulator — the program that executes the testbench — maintains the simulator clock, which is the master reference for all events in the testbench simulation. Events occur only at the instants dictated by the testbench HDL (such as a reset-toggle coded into the testbench), or in reaction (by the model) to stimulus and triggering events. Modern HDL simulators have full-featured graphical user interfaces, complete with a suite of debug tools. These allow the user to stop and restart the simulation at any time, insert simulator breakpoints (independent of the HDL code), and monitor or modify any element in the HDL model hierarchy. Modern simulators can also link the HDL environment to user-compiled libraries, through a defined PLI/VHPI interface. Linking is system-dependent (Win32/Linux/SPARC), as the HDL simulator and user libraries are compiled and linked outside the HDL environment.

Design verification is often the most time-consuming portion of the design process, due to the disconnect between a device"s functional specification, the designer"s interpretation of the specification, and the imprecision of the HDL language. The majority of the initial test/debug cycle is conducted in the HDL *simulator* environment, as the early stage of the design is subject to frequent and major circuit changes. An HDL description can also be prototyped and tested in hardware — programmable logic devices are often used for this purpose. Hardware prototyping is comparatively more expensive than HDL simulation, but offers a real-world view of the design. Prototyping is the best way to check interfacing against other hardware devices and hardware prototypes. Even those running on slow FPGAs offer much shorter simulation times than pure HDL simulation.

Design Verification with HDLs

Historically, design verification was a laborious, repetitive loop of writing and running simulation test cases against the design under test. As chip designs have grown larger and more complex, the task of design verification has grown to the point where it now dominates the schedule of a design team. Looking for ways to improve design productivity, the electronic design automation industry developed the Property Specification Language.

In formal verification terms, a property is a factual statement about the expected or assumed behavior of another object. Ideally, for a given HDL description, a property or properties can be proven true or false using formal mathematical methods. In practical terms, many properties cannot be proven because they occupy an unbounded solution space. However, if provided a set of operating assumptions or constraints, a property checker can prove (or disprove) certain properties by narrowing the solution space.

The assertions do not model circuit activity, but capture and document the designer"s intent in the HDL code. In a simulation environment, the simulator evaluates all specified assertions, reporting the location and severity of any violations. In a synthesis environment, the synthesis tool usually operates with the policy of halting synthesis upon any violation. Assertion-based verification is still in its infancy, but is expected to become an integral part of the HDL design toolset.

HDL and Programming Languages

A HDL is grossly similar to a software programming language, but there are major differences. Many programming languages are inherently procedural (single-threaded), with limited syntactical and semantic support to handle concurrency. HDLs, on the other hand, resemble concurrent programming languages in their ability to model multiple parallel processes (such as flipflops and adders) that automatically execute independently of one another. Any change to the process"s input automatically triggers an update in the simulator"s process stack. Both programming languages and HDLs are processed by a compiler (usually called a synthesizer in the HDL case), but with different goals. For HDLs, "compiling" refers to logical synthesis, a process of transforming the HDL code listing into a physically realizable gate netlist. The netlist output can take any of many forms: a "simulation" netlist with gate-delay information, a "handoff" netlist for post-synthesis place and route, or a generic industry-standard Electronic Design Interchange Format (EDIF) (for subsequent conversion to a JEDEC-format file).

On the other hand, a software compiler converts the source-code listing into a microprocessor-specific object code for execution on the target microprocessor. As HDLs and programming languages borrow concepts and features from each other, the boundary between them is becoming less distinct. However, pure HDLs are unsuitable for general purpose application software development, just as general-purpose programming languages are undesirable for modeling hardware. Yet as electronic systems grow increasingly complex, and reconfigurable systems become increasingly common, there is growing desire in the industry for a single language that can perform some tasks of both hardware design and software programming. SystemC is an example of such—embedded system hardware can be modeled as non-detailed architectural blocks (black boxes with modeled signal inputs and output drivers). The target application is written in C or C++ and natively compiled for the host-development system (as opposed to targeting the embedded CPU, which requires host-simulation of the embedded CPU or an emulated CPU). The high level of abstraction of SystemC models is well suited to early architecture exploration, as architectural modifications can be easily evaluated with little concern for signal-level implementation issues. However, the threading model used in SystemC and its reliance on shared memory mean that it does not handle parallel execution or lower level models well.

High-level Synthesis

In their level of abstraction, HDLs have been compared to assembly languages. There are moves to raise the abstraction level of the design in order to reduce the complexity of programming in HDLs, creating a sub-field called *high-level synthesis*. Companies such as Cadence, Synopsys and Agility Design Solutions are promoting SystemC as a way to combine high-level languages with concurrency models to allow faster design cycles for FPGAs than is possible using traditional HDLs. Approaches based on standard C or C++ (with libraries or other extensions allowing parallel programming) are found in the Catapult C tools from Mentor Graphics, and the Impulse C tools from Impulse Accelerated Technologies. Annapolis Micro Systems, Inc."s CoreFire Design Suite and National Instruments LabVIEW FPGA provide a graphical dataflow approach to high-level design entry and languages such as SystemVerilog, SystemVHDL, and Handel-C seek to accomplish the same goal, but are aimed at making existing hardware engineers more productive, rather than making FPGAs more accessible to existing software engineers. It is also possible to design hardware modules using MATLAB and Simulink using the Mathworks HDL Coder tool or Xilinx System Generator (XSG) (formerly Accel DSP) from Xilinx.

Examples of HDLs

HDLs for Analog Circuit Design

Name	Description
Analog Hardware Descriptive Language	an open analog hardware description language
SpectreHDL	a proprietary analog hardware description language
Verilog-AMS (Verilog for Analog and Mixed-Signal)	an open standard extending Verilog for analog and mixed analog/digital simulation
VHDL-AMS (VHDL with Analog/ Mixed-Signal extension)	a standardised language for mixed analog/digital simulation
HDL-A	a proprietary analog hardware description language

HDLs for Digital Circuit Design

The two most widely used and well-supported HDL varieties used in industry are Verilog and VHDL.

Name	Description
Advanced Boolean Expression Language (ABEL)	
Altera Hardware Description Language (AHDL)	a proprietary language from Altera
AHPL	A Hardware Programming language
Bluespec	high-level HDL based on Haskell (not embedded DSL)
Bluespec SystemVerilog (BSV)	based on Bluespec, with Verilog HDL like syntax, by Bluespec, Inc.
C-to-Verilog	Converter from C to Verilog
Chisel (Constructing Hardware in a Scala Embedded Language)	based on Scala (embedded DSL)
Confluence	a functional HDL; has been discontinued
CoWareC	a C-based HDL by CoWare. Now discontinued in favor of SystemC
CUPL (Compiler for Universal Programmable Logic)	a proprietary language from Logical Devices, Inc.
ELLA	no longer in common use
ESys.net	.net framework written in C#
Handel-C	a C-like design language
HHDL	based on Haskell (embedded DSL).
Hardware Join Java (HJJ)	based on Join Java
HML (Hardware ML)	based on SML
Hydra	based on Haskell
Impulse C	another C-like HDL
ISPS	Original HDL from CMU, no longer in common use
ParC (Parallel C++)	kusu extended with HDL style threading and communication for task-parallel programming
JHDL	based on Java
KARL	KARlsruhe Language (chapter in), a Pascalish hardware descriptive language, no longer in common use.
Lava	based on Haskell (embedded DSL).
Lola	a simple language used for teaching
M	A HDL from Mentor Graphics
MyHDL	based on Python (embedded DSL)
PALASM	for Programmable Array Logic (PAL) devices
ROCCC (Riverside Optimizing Compiler for Configurable Computing)	Free and open-source C to HDL tool
RHDL	based on the Ruby programming language

Ruby (hardware description language)	
SystemC	a standardized class of C++ libraries for high-level behavioral and transaction modeling of digital hardware at a high level of abstraction, i.e. system-level
SystemVerilog	a superset of Verilog, with enhancements to address system-level design and verification
SystemTCL	SDL based on Tcl.
THDL++ (Templated HDL inspired by C++)	An extension of VHDL with inheritance, advanced templates and policy classes
TL-Verilog (Transaction-Level Verilog)	An extension of Verilog/SystemVerilog with constructs for pipelines and transactions.
Verilog	One of the most widely used and well-supported HDLs
VHDL (VHSIC HDL)	One of the most widely used and well-supported HDLs

HDLs for printed Circuit Board Design

Several projects exist for defining printed circuit board connectivity using language based, textual-entry methods.

Name	**Description**
PHDL (PCB HDL)	A free and open source HDL for defining printed circuit board connectivity
EDAsolver	An HDL for solving schematic designs based on constraints

STATIC RANDOM-ACCESS MEMORY

Static random-access memory (**static RAM** or **SRAM**) is a type of semiconductor memory that uses bistable latching circuitry (flip-flop) to store each bit. SRAM exhibits data remanence, but it is still *volatile* in the conventional sense that data is eventually lost when the memory is not powered.

The term *static* differentiates SRAM from DRAM (*dynamic* random-access memory) which must be periodically refreshed. SRAM is faster and more expensive than DRAM; it is typically used for CPU cache while DRAM is used for a computer's main memory.

Applications and Uses

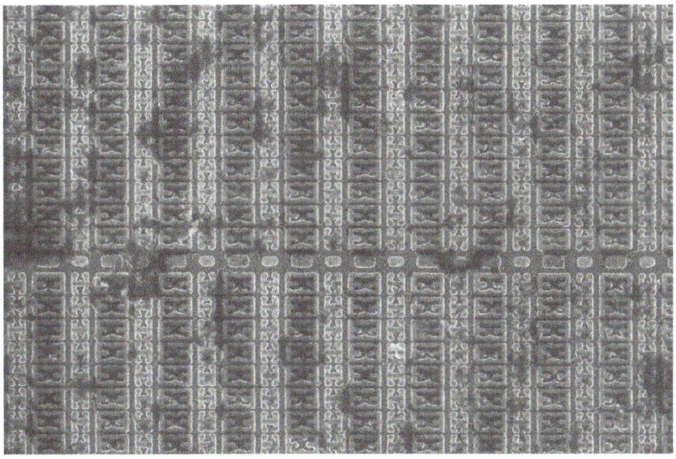

SRAM cells on the die of a STM32F103VGT6 microcontroller as seen by a scanning electron microscope. Manufactured by STMicroelectronics using a 180 nanometre process.

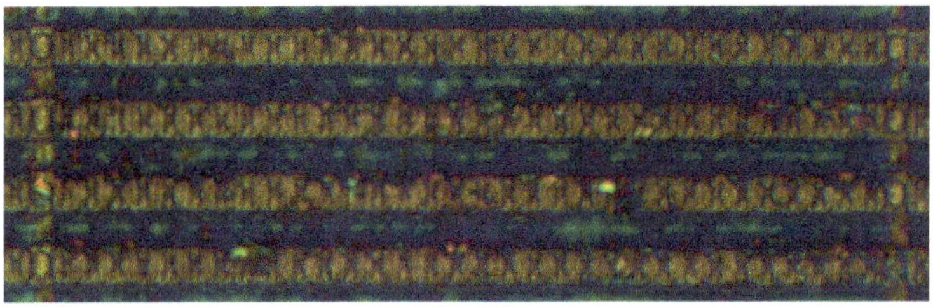

Comparison image of 180 nanometre SRAM cells on a STM32F103VGT6 microcontroller as seen by an optical microscope.

Characteristics

Advantages:

- Low power consumption
- Simplicity—a refresh circuit is not needed
- Reliability

Disadvantages:

- Price
- Capacity

Clock Rate and Power

The power consumption of SRAM varies widely depending on how frequently it is accessed; in some instances, it can use as much power as dynamic RAM, when used at high frequencies, and some ICs can consume many watts at full bandwidth. On the other hand, static RAM used at a somewhat slower pace, such as in applications with moderately clocked microprocessors, draws very little power and can have a nearly negligible power consumption when sitting idle – in the region of a few micro-watts. Several techniques have been proposed to manage power consumption of SRAM-based memory structures.

STATIC RAM exists Primarily as:

- general purpose products.
  - with *asynchronous* interface, such as the ubiquitous 28-pin 8K × 8 and 32K × 8 chips (often but not always named something along the lines of 6264 and 62C256 respectively), as well as similar products up to 16 Mbit per chip
  - with *synchronous* interface, usually used for caches and other applications requiring burst transfers, up to 18 Mbit (256K × 72) per chip
- integrated on chip
  - as RAM or cache memory in micro-controllers (usually from around 32 bytes up to 128 kilobytes)
  - as the primary caches in powerful microprocessors, such as the x86 family, and many others (from 8 KB, up to many megabytes)
  - to store the registers and parts of the state-machines used in some microprocessors (see register file)
  - on application specific ICs, or ASICs (usually in the order of kilobytes)
  - in FPGAs and CPLDs

Embedded use

- Many categories of industrial and scientific subsystems, automotive electronics, and similar, contain static RAM.
- Some amount (kilobytes or less) is also embedded in practically all modern appliances, toys, etc. that implement an electronic user interface.
- Several megabytes may be used in complex products such as digital cameras, cell phones, synthesizers, etc.

SRAM in its dual-ported form is sometimes used for realtime digital signal processing circuits.

In computers

SRAM is also used in personal computers, workstations, routers and peripheral equipment: CPU register files, internal CPU caches and external burst mode SRAM caches, hard disk buffers, router buffers, etc. LCD screens and printers also normally employ static RAM to hold the image displayed (or to be printed). Static RAM was used for the main memory of some early personal computers such as the ZX80, TRS-80 Model 100 and Commodore VIC-20.

Hobbyists

Hobbyists, specifically homebuilt processor enthusiasts, often prefer SRAM due to the ease of interfacing. It is much easier to work with than DRAM as there are no refresh cycles and the address and data buses are directly accessible rather than multiplexed. In addition to buses and power connections, SRAM usually requires only three controls: Chip Enable (CE), Write Enable (WE) and Output Enable (OE). In synchronous SRAM, Clock (CLK) is also included.

Types of SRAM

Non-volatile SRAM

Non-volatile SRAMs, or nvSRAMs, have standard SRAM functionality, but they save the data when the power supply is lost, ensuring preservation of critical information. nvSRAMs are used in a wide range of situations—networking, aerospace, and medical, among many others —where the preservation of data is critical and where batteries are impractical.

By transistor type

- Bipolar junction transistor (used in TTL and ECL) – very fast but consumes a lot of power
- MOSFET (used in CMOS) – low power and very common today

By function

- Asynchronous – independent of clock frequency; data in and data out are controlled by address transition
- Synchronous – all timings are initiated by the clock edge(s). Address, data in and other control signals are associated with the clock signals

In 1990s asynchronous SRAM used to be employed for fast access time. Asynchronous SRAM was used as main memory for small cache-less embedded processors used in everything from industrial electronics and measurement systems to hard disks and networking equipment, among many other applications. Nowadays, synchronous SRAM (e.g. DDR SRAM) is rather employed similarly like Synchronous DRAM - DDR SDRAM memory is rather used than asynchronous DRAM (Dynamic random-access memory). Synchronous memory interface is much faster as access time can be significantly reduced by employing pipeline architecture. Furthermore, as DRAM is much cheaper than SRAM, SRAM is often replaced by DRAM, especially in the case when large volume of data is required. SRAM memory is however much faster for random (not block / burst) access. Therefore, SRAM memory is mainly used for CPU cache, small on-chip memory, FIFOs or other small buffers.

By feature

- ZBT (ZBT stands for zero bus turnaround) – the turnaround is the number of clock cycles it takes to change access to the SRAM from *write* to *read* and vice versa. The turnaround for ZBT SRAMs or the latency between read and write cycle is zero.
- syncBurst (syncBurst SRAM or synchronous-burst SRAM) – features synchronous burst write access to the SRAM to increase write operation to the SRAM
- DDR SRAM – Synchronous, single read/write port, double data rate I/O
- Quad Data Rate SRAM – Synchronous, separate read and write ports, quadruple data rate I/O

By flip-flop type

- Binary SRAM
- Ternary SRAM

Design

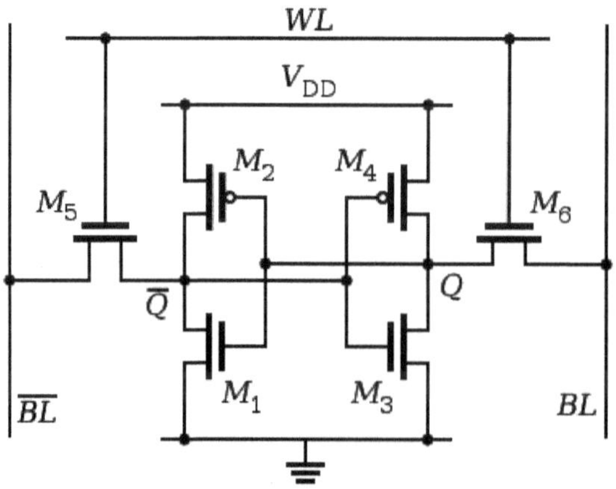

A six-transistor CMOS SRAM cell.

A typical SRAM cell is made up of six MOSFETs. Each bit in an SRAM is stored on four transistors (M1, M2, M3, M4) that form two cross-coupled inverters. This storage cell has two stable states which are used to denote **0** and **1**. Two additional *access* transistors serve to control the access to a storage cell during read and write operations. In addition to such six-transistor (6T) SRAM, other kinds of SRAM chips use 4, 8, 10 (4T, 8T, 10T SRAM), or more transistors per bit. Four-transistor SRAM is quite common in stand-alone SRAM devices (as opposed to SRAM used for CPU caches), implemented in special processes with an extra layer of polysilicon, allowing for very high-resistance pull-up resistors. The principal drawback of using 4T SRAM is increased static power due to the constant current flow through one of the pull-down transistors.

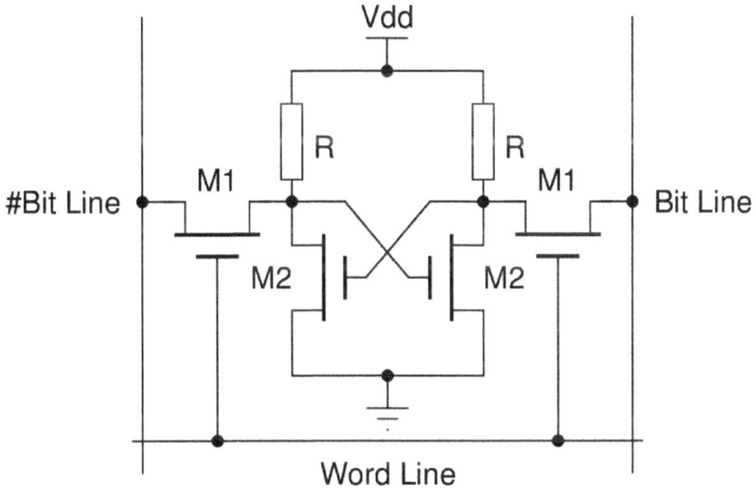

Four transistor SRAM provides advantages in density at the cost of manufacturing complexity. The resistors must have small dimensions and large values.

This is sometimes used to implement more than one (read and/or write) port, which may be useful in certain types of video memory and register files implemented with multi-ported SRAM circuitry.

Generally, the fewer transistors needed per cell, the smaller each cell can be. Since the cost of processing a silicon wafer is relatively fixed, using smaller cells and so packing more bits on one wafer reduces the cost per bit of memory.

Memory cells that use fewer than four transistors are possible – but, such 3T or 1T cells are DRAM, not SRAM (even the so-called 1T-SRAM).

Access to the cell is enabled by the word line (WL in figure) which controls the two *access* transistors M_5 and M_6 which, in turn, control whether the cell should be connected to the bit lines: BL and BL. They are used to transfer data for both read and write operations. Although it is not strictly necessary to have two bit lines, both the signal and its inverse are typically provided in order to improve noise margins.

During read accesses, the bit lines are actively driven high and low by the inverters in the SRAM cell. This improves SRAM bandwidth compared to DRAMs – in a DRAM, the bit line is connected to storage capacitors and charge sharing causes the bitline to swing upwards or downwards. The symmetric structure of SRAMs also allows for differential signaling, which makes small voltage swings more easily detectable. Another difference with DRAM that contributes to making SRAM faster is that commercial chips accept all address bits at a time. By comparison, commodity DRAMs have the address multiplexed in two halves, i.e. higher bits followed by lower bits, over the same package pins in order to keep their size and cost down.

The size of an SRAM with m address lines and n data lines is 2^m words, or $2^m \times n$ bits. The most common word size is 8 bits, meaning that a single byte can be read or written to each of 2^m different words within the SRAM chip. Several common SRAM chips have 11 address lines (thus a capacity of $2^m = 2{,}048 = 2\text{k}$ words) and an 8-bit word, so they are referred to as "2k × 8 SRAM".

SRAM Operation

An SRAM cell has three different states: *standby* (the circuit is idle), *reading* (the data has been requested) or *writing* (updating the contents). SRAM operating in read mode and write modes should have "readability" and "write stability", respectively. The three different states work as follows:

Standby

If the word line is not asserted, the *access* transistors M_5 and M_6 disconnect the cell from the bit lines. The two cross-coupled inverters formed by $M_1 - M_4$ will continue to reinforce each other as long as they are connected to the supply.

Reading

In theory, reading only requires asserting the word line WL and reading the SRAM cell state by a single access transistor and bit line, e.g. M_6, BL. Nevertheless, bit lines are relatively long and have large parasitic capacitance. To speed up reading, a more complex process is used in practice: The read cycle is started by precharging both bit lines BL and BL, i.e., driving the bit lines to a threshold voltage (midrange voltage between logical **1** and **0**) by an external module (not shown in the figures). Then asserting the word line WL enables both the access transistors M_5 and M_6, which causes the bit line BL voltage to either slightly drop (bottom NMOS transistor M_3 is ON and top PMOS transistor M_4 is off) or rise (top PMOS transistor M_4 is on). It should be noted that if BL voltage rises, the BL voltage drops, and vice versa. Then the BL and BL lines will have a small voltage difference between them. A sense amplifier will sense which line has the higher voltage and thus determine whether there was **1** or **0** stored. The higher the sensitivity of the sense amplifier, the faster the read operation.

Writing

The write cycle begins by applying the value to be written to the bit lines. If we wish to write a **0**, we would apply a **0** to the bit lines, i.e. setting BL to **1** and BL to **0**. This is similar to applying a reset pulse to an SR-latch, which causes the flip flop to change state. A **1** is written by inverting the values of the bit lines. WL is then asserted and the value that is to be stored is latched in. This works because the bit line input-drivers are designed to be much stronger than the relatively weak transistors in the cell itself so they can easily override the previous state of the cross-coupled inverters. In practice, access NMOS transistors M_5 and M_6 have to be stronger than either bottom NMOS (M_1, M_3) or top PMOS (M_2, M_4) transistors. This is easily obtained as PMOS transistors are much weaker than NMOS when same sized. Consequently when one transistor pair (e.g. M_3 and M_4) is only slightly overriden by the write process, the opposite transistors pair (M_1 and M_2) gate voltage is also changed. This means that the M_1 and M_2 transistors can be easier overriden, and so on. Thus, cross-coupled inverters magnify the writing process.

Bus Behavior

RAM with an access time of 70 ns will output valid data within 70 ns from the time that the address lines are valid. But the data will remain for a hold time as well (5–10 ns). Rise and fall times also influence valid timeslots with approximately 5 ns. By reading the lower part of an address range, bits in sequence (page cycle) one can read with significantly shorter access time (30 ns).

FLASH MEMORY

Flash memory is an electronic (solid-state) non-volatile computer storage medium that can be electrically erased and reprogrammed.

Toshiba developed flash memory from EEPROM (electrically erasable programmable read-only memory) in the early 1980s and introduced it to the market in 1984. The two main types of flash memory are named after the NAND and NOR logic gates. The individual flash memory cells exhibit internal characteristics similar to those of the corresponding gates.

A USB flash drive. The chip on the left is the flash memory. The controller is on the right.

Whereas EPROMs had to be completely erased before being rewritten, NAND-type flash memory may be written and read in blocks (or pages) which are generally much smaller than the entire device. NOR-type flash allows a single machine word (byte) to be written—to an erased location—or read independently.

The NAND type operates primarily in memory cards, USB flash drives, solid-state drives (those produced in 2009 or later), and similar products, for general storage and transfer of data. NAND or NOR flash memory is also often used to store configuration data in numerous digital products, a task previously made possible by EEPROM or battery-powered static RAM. One key disadvantage of flash memory is that it can only endure a relatively small number of write cycles in a specific block.

Example applications of both types of flash memory include personal computers, PDAs, digital audio players, digital cameras, mobile phones, synthesizers, video games, scientific instrumentation, industrial robotics, and medical electronics. In addition to being non-volatile, flash memory offers fast read access times, although not as fast as static RAM or ROM. Its mechanical shock resistance helps explain its popularity over hard disks in portable devices, as does its high durability, ability to withstand high pressure, temperature and immersion in water, etc.

Although flash memory is technically a type of EEPROM, the term "EEPROM" is generally used to refer specifically to non-flash EEPROM which is erasable in small blocks, typically bytes. Because erase cycles are slow, the large block sizes used in flash memory erasing give it a significant speed advantage over non-flash EEPROM when writing large amounts of data. As of 2013, flash memory costs much less than byte-programmable EEPROM and had become the dominant memory type wherever a system required a significant amount of non-volatile solid-state storage.

Flash memory (both NOR and NAND types) was invented by Fujio Masuoka while working for Toshiba circa 1980. According to Toshiba, the name "flash" was suggested by Masuoka's colleague, Shōji Ariizumi, because the erasure process of the memory contents reminded him of the flash of a camera. Masuoka and colleagues presented the invention at the *IEEE 1984 International Electron Devices Meeting* (IEDM) held in San Francisco.

Intel Corporation saw the massive potential of the invention and introduced the first commercial NOR type flash chip in 1988. NOR-based flash has long erase and write times, but provides full address and data buses, allowing random access to any memory location. This makes it a suitable replacement for older read-only memory (ROM) chips, which are used to store program code that rarely needs to be updated, such as a computer's BIOS or the firmware of set-top boxes. Its endurance may be from as little as 100 erase cycles for an on-chip flash memory, to a more typical 10,000 or 100,000 erase cycles, up to 1,000,000 erase cycles. NOR-based flash was the basis of early flash-based removable media; CompactFlash was originally based on it, though later cards moved to less expensive NAND flash.

NAND flash has reduced erase and write times, and requires less chip area per cell, thus allowing greater storage density and lower cost per bit than NOR flash; it also has up to 10 times the endurance of NOR flash. However, the I/O interface of NAND flash does not provide a random-access external address bus. Rather, data must be read on a block-wise basis, with typical block sizes of hundreds to thousands of bits. This makes NAND flash unsuitable as a drop-in replacement for program ROM, since most microprocessors and microcontrollers require byte-level random access. In this regard, NAND flash is similar to other secondary data storage devices, such as hard disks and optical media, and is thus, highly suitable for use in mass-storage devices, such as memory cards. The first NAND-based removable media format was SmartMedia in 1995, and many others have followed, including:

- MultiMediaCard
- Secure Digital

- Memory Stick, and xD-Picture Card.

A new generation of memory card formats, including RS-MMC, miniSD and microSD, feature extremely small form factors. For example, the microSD card has an area of just over 1.5 cm^2, with a thickness of less than 1 mm. microSD capacities range from 64 MB to 256 GB, as of May 2016.

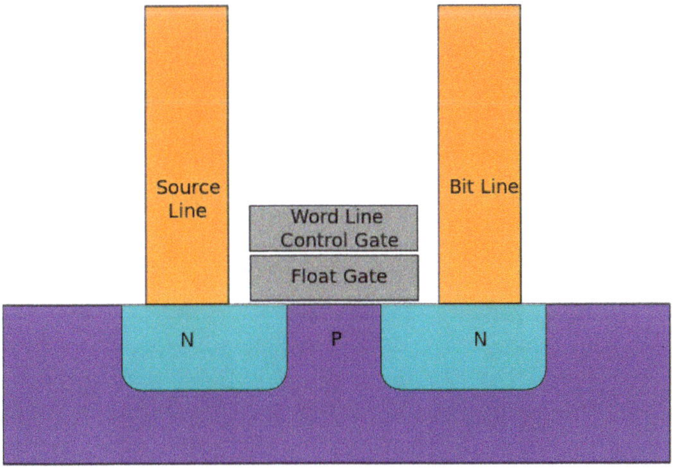

A flash memory cell.

Principles of Operation

Flash memory stores information in an array of memory cells made from floating-gate transistors. In single-level cell (SLC) devices, each cell stores only one bit of information. In multi-level cell (MLC) devices, including triple-level cell (TLC) devices, can store more than one bit per cell.

The floating gate may be conductive (typically polysilicon in most kinds of flash memory) or non-conductive (as in SONOS flash memory).

Floating-Gate Transistor

In flash memory, each memory cell resembles a standard MOSFET, except that the transistor has two gates instead of one. On top is the control gate (CG), as in other MOS transistors, but below this there is a floating gate (FG) insulated all around by an oxide layer. The FG is interposed between the CG and the MOSFET channel. Because the FG is electrically isolated by its insulating layer, electrons placed on it are trapped until they are removed by another application of electric field (e.g. Applied voltage or UV as in EPROM). Counter-intuitively, placing electrons on the FG sets the transistor to the logical "0" state. Once the FG is charged, the electrons in it screen (partially cancel) the electric field from the CG, thus, increasing the threshold voltage (V_{T1}) of the cell. This means that now a higher voltage(V_{T2}) must be applied to the CG to make the channel conductive. In order to read a value from the transistor, an intermediate voltage between the threshold voltages (V_{T1} & V_{T2}) is applied to the CG. If the channel conducts at this intermediate voltage, the FG must be uncharged (if it were charged, we would not get conduction because the intermediate voltage is less than V_{T2}), and hence, a logical "1" is stored in the gate. If the channel does not conduct at the intermediate voltage, it indicates that the FG is charged, and hence, a logical "0" is stored in the gate. The presence of a logical "0" or "1" is sensed by determining whether there is current flowing through the transistor when the intermediate voltage is asserted on the CG. In a multi-level cell device, which stores more than one bit per cell, the amount of current flow is sensed (rather than simply its presence or absence), in order to determine more precisely the level of charge on the FG.

Internal Charge Pumps

Despite the need for high programming and erasing voltages, virtually all flash chips today require only a single supply voltage, and produce the high voltages using on-chip charge pumps.

Over half the energy used by a 1.8 V NAND flash chip is lost in the charge pump itself. Since boost converters are inherently more efficient than charge pumps, researchers developing low-power SSDs have proposed returning to the dual Vcc/Vpp supply voltages used on all the early flash chips, driving the high Vpp voltage for all flash chips in a SSD with a single shared external boost converter.

In spacecraft and other high-radiation environments, the on-chip charge pump is the first part of the flash chip to fail, although flash memories will continue to work—in read-only mode—at much higher radiation levels.

NOR Flash

In NOR flash, each cell has one end connected directly to ground, and the other end connected directly to a bit line. This arrangement is called "NOR flash" because it acts like a NOR gate: when one of the word lines (connected to the cell's CG) is brought high, the corresponding storage transistor acts to pull the output bit line low. NOR flash continues to be the technology of choice for embedded applications requiring a discrete non-volatile memory device. The low read latencies characteristic of NOR devices allow for both direct code execution and data storage in a single memory product.

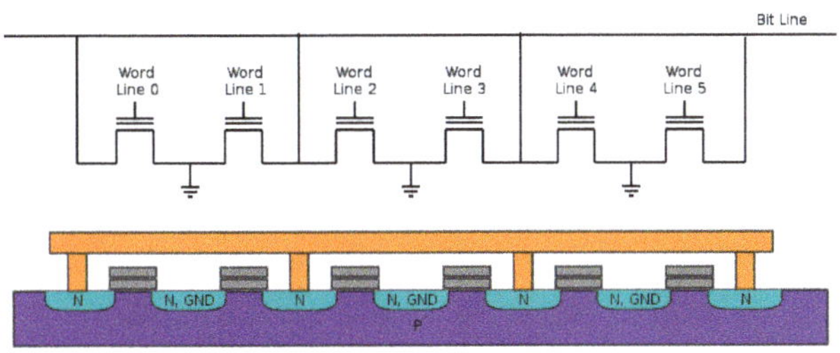

NOR flash memory wiring and structure on silicon

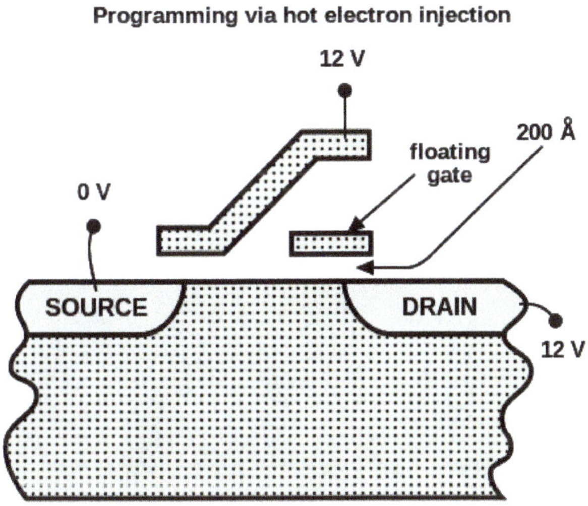

Programming a NOR memory cell (setting it to logical 0), via hot-electron injection.

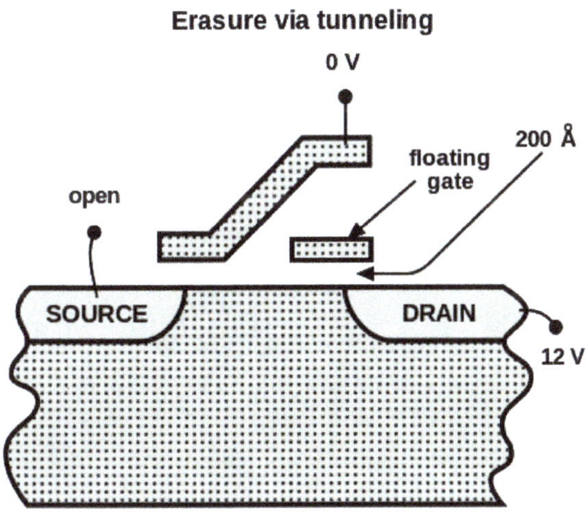

Erasing a NOR memory cell (setting it to logical 1), via quantum tunneling.

Programming

A single-level NOR flash cell in its default state is logically equivalent to a binary "1" value, because current will flow through the channel under application of an appropriate voltage to the control gate, so that the bitline voltage is pulled down. A NOR flash cell can be programmed, or set to a binary "0" value, by the following procedure:

- an elevated on-voltage (typically >5 V) is applied to the CG
- the channel is now turned on, so electrons can flow from the source to the drain (assuming an NMOS transistor)
- the source-drain current is sufficiently high to cause some high energy electrons to jump through the insulating layer onto the FG, via a process called hot-electron injection.

Erasing

To erase a NOR flash cell (resetting it to the "1" state), a large voltage *of the opposite polarity* is applied between the CG and source terminal, pulling the electrons off the FG through quantum tunneling. Modern NOR flash memory chips are divided into erase segments (often called blocks or sectors). The erase operation can be performed only on a block-wise basis; all the cells in an erase segment must be erased together. Programming of NOR cells, however, generally can be performed one byte or word at a time.

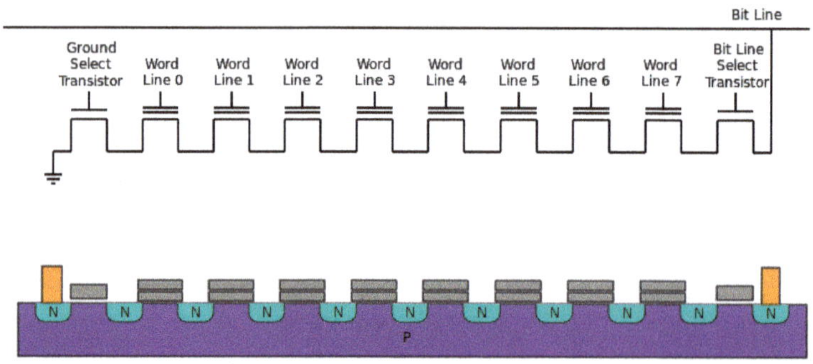

NAND flash memory wiring and structure on silicon

NAND Flash

NAND flash also uses floating-gate transistors, but they are connected in a way that resembles a NAND gate: several transistors are connected in series, and the bit line is pulled low only if all the word lines are pulled high (above the transistors' V_T). These groups are then connected via some additional transistors to a NOR-style bit line array in the same way that single transistors are linked in NOR flash.

Compared to NOR flash, replacing single transistors with serial-linked groups adds an extra level of addressing. Whereas NOR flash might address memory by page then word, NAND flash might address it by page, word and bit. Bit-level addressing suits bit-serial applications (such as hard disk emulation), which access only one bit at a time. Execute-in-place applications, on the other hand, require every bit in a word to be accessed simultaneously. This requires word-level addressing. In any case, both bit and word addressing modes are possible with either NOR or NAND flash.

To read data, first the desired group is selected (in the same way that a single transistor is selected from a NOR array). Next, most of the word lines are pulled up above the V_T of a programmed bit, while one of them is pulled up to just over the V_T of an erased bit. The series group will conduct (and pull the bit line low) if the selected bit has not been programmed.

Despite the additional transistors, the reduction in ground wires and bit lines allows a denser layout and greater storage capacity per chip. (The ground wires and bit lines are actually much wider than the lines in the diagrams.) In addition, NAND flash is typically permitted to contain a certain number of faults (NOR flash, as is used for a BIOS ROM, is expected to be fault-free). Manufacturers try to maximize the amount of usable storage by shrinking the size of the transistors.

Writing and erasing

NAND flash uses tunnel injection for writing and tunnel release for erasing. NAND flash memory forms the core of the removable USB storage devices known as USB flash drives, as well as most memory card formats and solid-state drives available today.

Vertical NAND

Vertical NAND (V-NAND) memory stacks memory cells vertically and uses a charge trap flash architecture. The vertical layers allow larger areal bit densities without requiring smaller individual cells.

Structure

V-NAND uses a charge trap flash geometry (pioneered in 2002 by AMD) that stores charge on an embedded silicon nitride film. Such a film is more robust against point defects and can be made thicker to hold larger numbers of electrons. V-NAND wraps a planar charge trap cell into a cylindrical form.

An individual memory cell is made up of one planar polysilicon layer containing a hole filled by multiple concentric vertical cylinders. The hole"s polysilicon surface acts as the gate electrode. The outermost silicon dioxide cylinder acts as the gate dielectric, enclosing a silicon nitride cylinder that stores charge, in turn enclosing a silicon dioxide cylinder as the tunnel dielectric that surrounds a central rod of conducting polysilicon which acts as the conducting channel.

Memory cells in different vertical layers do not interfere with each other, as the charges cannot move vertically through the silicon nitride storage medium, and the electric fields associated with the gates are closely confined within each layer. The vertical collection is electrically identical to the serial-linked groups in which conventional NAND flash memory is configured.

Construction

Growth of a group of V-NAND cells begins with an alternating stack of conducting (doped) polysilicon layers and insulating silicon dioxide layers.

The next step is to form a cylindrical hole through these layers. In practice, a 128 Gibit V-NAND chip with 24 layers of memory cells requires about 2.9 billion such holes. Next the hole's inner surface receives multiple coatings, first silicon dioxide, then silicon nitride, then a second layer of silicon dioxide. Finally, the hole is filled with conducting (doped) polysilicon.

Performance

As of 2013, V-NAND flash architecture allows read and write operations twice as fast as conventional NAND and can last up to 10 times as long, while consuming 50 percent less power. They offer comparable physical bit density using 10-nm lithography, but may be able to increase bit density by up to two orders of magnitude.

Limitations

Block Erasure

One limitation of flash memory is that, although it can be read or programmed a byte or a word at a time in a random access fashion, it can be erased only a block at a time. This generally sets all bits in the block to 1. Starting with a freshly erased block, any location within that block can be programmed. However, once a bit has been set to 0, only by erasing the entire block can it be changed back to 1. In other words, flash memory (specifically NOR flash) offers random-access read and programming operations, but does not offer arbitrary random-access rewrite or erase operations. A location can, however, be rewritten as long as the new value's 0 bits are a superset of the over-written values. For example, a nibble value may be erased to 1111, then written as 1110. Successive writes to that nibble can change it to 1010, then 0010, and finally 0000. Essentially, erasure sets all bits to 1, and programming can only clear bits to 0. File systems designed for flash devices can make use of this capability, for example, to represent sector metadata.

Although data structures in flash memory cannot be updated in completely general ways, this allows members to be "removed" by marking them as invalid. This technique may need to be modified for multi-level cell devices, where one memory cell holds more than one bit.

Common flash devices such as USB flash drives and memory cards provide only a block-level interface, or flash translation layer (FTL), which writes to a different cell each time to wear-level the device. This prevents incremental writing within a block; however, it does not help the device from being prematurely worn out by intensive write patterns.

Memory Wear

Another limitation is that flash memory has a finite number of program–erase cycles (typically written as P/E cycles). Most commercially available flash products are guaranteed to withstand around 100,000 P/E cycles before the wear begins to deteriorate the integrity of the storage. Micron Technology and Sun Microsystems announced an SLC NAND flash memory chip rated for 1,000,000 P/E cycles on 17 December 2008.

The guaranteed cycle count may apply only to block zero (as is the case with TSOP NAND devices), or to all blocks (as in NOR). This effect is mitigated in some chip firmware or file system drivers by counting the writes and dynamically remapping blocks in order to spread write operations between sectors; this technique is called wear leveling. Another approach is to perform write verification and remapping to spare sectors in case of write failure, a technique called bad block management (BBM). For portable consumer devices, these wearout management techniques typically extend the life of the flash memory beyond the life of the device itself, and some data loss may be acceptable in these applications. For high reliability data storage, however, it is not advisable to use flash memory that would have to go through a large number of programming cycles. This limitation is meaningless for 'read-only' applications such as thin clients and routers, which are programmed only once or at most a few times during their lifetimes.

In December 2012, Taiwanese engineers from Macronix revealed their intention to announce at the 2012 IEEE International Electron Devices Meeting that it has figured out how to improve NAND flash storage read/write cycles from 10,000 to 100 million cycles using a "self-healing" process that uses a flash chip with "onboard heaters that could anneal small groups of memory cells." The built-in thermal annealing replaces the usual erase cycle with a local high temperature process that not only erases the stored charge, but also repairs the electron-induced stress in the chip, giving write cycles of at least 100 million. The result is a chip that can be erased and rewritten over and over, even when it should theoretically break down. As promising as Macronix's breakthrough could be for the mobile industry, however, there are no plans for a commercial product to be released any time in the near future.

Read Disturb

The method used to read NAND flash memory can cause nearby cells in the same memory block to change over time (become programmed). This is known as read disturb. The threshold number of reads is generally in the hundreds of thousands of reads between intervening erase operations. If reading continually from one cell, that cell will not fail but rather one of the surrounding cells on a subsequent read. To avoid the read disturb problem the flash controller will typically count the total number of reads to a block since the last erase. When the count exceeds a target limit, the affected block is copied over to a new block, erased, then released to the block pool. The original block is as good as new after the erase. If the flash controller does not intervene in time, however, a read disturb error will occur with possible data loss if the errors are too numerous to correct with an error-correcting code.

X-ray Effects

Most flash ICs come in ball grid array (BGA) packages, and even the ones that do not are often mounted on a PCB next to other BGA packages. After PCB Assembly, boards with BGA packages are often X-rayed to see if the balls are making proper connections to the proper pad, or if the BGA needs rework. These X-rays can erase programmed bits in a flash chip (convert programmed "0" bits into erased "1" bits). Erased bits ("1" bits) are not affected by X-rays.

Some manufacturers are now making X-ray proof SD and USB memory devices.

Low-level access

The low-level interface to flash memory chips differs from those of other memory types such as DRAM, ROM, and EEPROM, which support bit-alterability (both zero to one and one to zero) and random access via externally accessible address buses.

NOR memory has an external address bus for reading and programming. For NOR memory, reading and programming are random-access, and unlocking and erasing are block-wise. For NAND memory, reading and programming are page-wise, and unlocking and erasing are block-wise.

NOR Memories

Reading from NOR flash is similar to reading from random-access memory, provided the address and data bus are mapped correctly. Because of this, most microprocessors can use NOR flash memory as execute in place (XIP) memory, meaning that programs stored in NOR flash can be executed directly from the NOR flash without needing to be copied into RAM first. NOR flash may be programmed in a random-access manner similar to reading. Programming changes bits from a logical one to a zero. Bits that are already zero are left unchanged. Erasure must happen a block at a time, and resets all the bits in the erased block back to one. Typical block sizes are 64, 128, or 256 KiB.

Bad block management is a relatively new feature in NOR chips. In older NOR devices not supporting bad block management, the software or device driver controlling the memory chip must correct for blocks that wear out, or the device will cease to work reliably.

The specific commands used to lock, unlock, program, or erase NOR memories differ for each manufacturer. To avoid needing unique driver software for every device made, special Common Flash Memory Interface (CFI) commands allow the device to identify itself and its critical operating parameters.

Besides its use as random-access ROM, NOR flash can also be used as a storage device, by taking advantage of random-access programming. Some devices offer read-while-write functionality so that code continues to execute even while a program or erase operation is occurring in the background. For sequential data writes, NOR flash chips typically have slow write speeds, compared with NAND flash.

Typical NOR flash does not need an error correcting code.

NAND memories

NAND flash architecture was introduced by Toshiba in 1989. These memories are accessed much like block devices, such as hard disks. Each block consists of a number of pages. The pages are typically 512 or 2,048 or 4,096 bytes in size. Associated with each page are a few bytes (typically 1/32 of the data size) that can be used for storage of an error correcting code (ECC) checksum.

Typical block sizes include:

- 32 pages of 512+16 bytes each for a block size of 16 kB
- 64 pages of 2,048+64 bytes each for a block size of 128 kB
- 64 pages of 4,096+128 bytes each for a block size of 256 kB
- 128 pages of 4,096+128 bytes each for a block size of 512 kB.

While reading and programming is performed on a page basis, erasure can only be performed on a block basis.

NAND devices also require bad block management by the device driver software, or by a separate controller chip. SD cards, for example, include controller circuitry to perform bad block management and wear leveling. When a logical block is accessed by high-level software, it is mapped to a physical block by the device driver or controller. A number of blocks on the flash chip may be set aside for storing mapping tables to deal with bad blocks, or the system may simply check each block at power-up to create a bad block map in RAM. The overall memory capacity gradually shrinks as more blocks are marked as bad.

NAND relies on ECC to compensate for bits that may spontaneously fail during normal device operation. A typical ECC will correct a one-bit error in each 2048 bits (256 bytes) using 22 bits of ECC, or a one-bit error in each 4096 bits (512 bytes) using 24 bits of ECC. If the ECC cannot correct the error during read, it may still detect the error. When doing erase or program operations, the device can detect blocks that fail to program or erase and mark them bad. The data is then written to a different, good block, and the bad block map is updated.

Hamming codes are the most commonly used ECC for SLC NAND flash. Reed-Solomon codes and Bose-Chaudhuri-Hocquenghem codes are commonly used ECC for MLC NAND flash. Some MLC NAND flash chips internally generate the appropriate BCH error correction codes.

Most NAND devices are shipped from the factory with some bad blocks. These are typically marked according to a specified bad block marking strategy. By allowing some bad blocks, the manufacturers achieve far higher yields than would be possible if all blocks had to be verified good. This significantly reduces NAND flash costs and only slightly decreases the storage capacity of the parts.

When executing software from NAND memories, virtual memory strategies are often used: memory contents must first be paged or copied into memory-mapped RAM and executed there (leading to the common combination of NAND + RAM). A memory management unit (MMU) in the system is helpful, but this can also be accomplished with overlays. For this reason, some systems will use a combination of NOR and NAND memories, where a smaller NOR memory is used as software ROM and a larger NAND memory is partitioned with a file system for use as a non-volatile data storage area.

NAND sacrifices the random-access and execute-in-place advantages of NOR. NAND is best suited to systems requiring high capacity data storage. It offers higher densities, larger capacities, and lower cost. It has faster erases, sequential writes, and sequential reads.

Standardization

A group called the Open NAND Flash Interface Working Group (ONFI) has developed a standardized low-level interface for NAND flash chips. This allows interoperability between conforming NAND devices from different vendors. The ONFI specification version 1.0 was released on 28 December 2006. It specifies:

- a standard physical interface (pinout) for NAND flash in TSOP-48, WSOP-48, LGA-52, and BGA-63 packages
- a standard command set for reading, writing, and erasing NAND flash chips
- a mechanism for self-identification (comparable to the serial presence detection feature of SDRAM memory modules)

The ONFI group is supported by major NAND flash manufacturers, including Hynix, Intel, Micron Technology, and Numonyx, as well as by major manufacturers of devices incorporating NAND flash chips.

Two major flash device manufacturers, Toshiba and Samsung, have chosen to use an interface of their own design known as Toggle Mode (and now Toggle V2.0). This interface isn"t pin-to-pin compatible with the ONFI specification. The result is a product designed for one vendor's devices may not be able to use another vendor's devices.

A group of vendors, including Intel, Dell, and Microsoft, formed a Non-Volatile Memory Host Controller Interface (NVMHCI) Working Group. The goal of the group is to provide standard software and hardware programming interfaces for nonvolatile memory subsystems, including the "flash cache" device connected to the PCI Express bus.

Distinction between NOR and NAND flash

NOR and NAND flash differ in two important ways:

- the connections of the individual memory cells are different
- the interface provided for reading and writing the memory is different (NOR allows random-access for reading, NAND allows only page access)

These two are linked by the design choices made in the development of NAND flash. A goal of NAND flash development was to reduce the chip area required to implement a given capacity of flash memory, and thereby to reduce cost per bit and increase maximum chip capacity so that flash memory could compete with magnetic storage devices like hard disks.

NOR and NAND flash get their names from the structure of the interconnections between memory cells. In NOR flash, cells are connected in parallel to the bit lines, allowing cells to be read and programmed individually. The parallel connection of cells resembles the parallel connection of transistors in a CMOS NOR gate. In NAND flash, cells are connected in series, resembling a NAND gate. The series connections consume less space than parallel ones, reducing the cost of NAND flash. It does not, by itself, prevent NAND cells from being read and programmed individually.

Each NOR flash cell is larger than a NAND flash cell — 10 F^2 vs 4 F^2 — even when using exactly the same semiconductor device fabrication and so each transistor, contact, etc. is exactly the same size—because NOR flash cells require a separate metal contact for each cell.

When NOR flash was developed, it was envisioned as a more economical and conveniently rewritable ROM than contemporary EPROM and EEPROM memories. Thus random-access reading circuitry was necessary. However, it was expected that NOR flash ROM would be read much more often than written, so the write circuitry included was fairly slow and could erase only in a block-wise fashion. On the other hand, applications that use flash as a replacement for disk drives do not require word-level write address, which would only add to the complexity and cost unnecessarily.

Because of the series connection and removal of wordline contacts, a large grid of NAND flash memory cells will occupy perhaps only 60% of the area of equivalent NOR cells(assuming the same CMOS process resolution, for example, 130 nm, 90 nm, or 65 nm). NAND flash"s designers realized that the area of a NAND chip, and thus the cost, could be further reduced by removing the external address and data bus circuitry. Instead, external devices could communicate with NAND flash via sequential-accessed command and data registers, which would internally retrieve and output the necessary data. This design choice made random-access of NAND flash memory impossible, but the goal of NAND flash was to replace mechanical hard disks, not to replace ROMs.

Attribute	**NAND**	**NOR**
Main Application	File storage	Code execution
Storage capacity	High	Low
Cost per bit	Better	
Active Power	Better	
Standby Power		Better
Write Speed	Good	
Read Speed		Good

Write Endurance

The write endurance of SLC floating-gate NOR flash is typically equal to or greater than that of NAND flash, while MLC NOR and NAND flash have similar endurance capabilities. Examples of endurance cycle ratings listed in datasheets for NAND and NOR flash are provided.

Type of flash memory	Endurance rating (Erases per block)	Example(s) of flash memory
SLC NAND	100,000	Samsung OneNAND KFW4G16Q2M
MLC NAND	5,000 to 10,000 for medium-capacity applications; 1,000 to 3,000 for high-capacity applications	Samsung K9G8G08U0M (Example for medium-capacity applications)
TLC NAND	1,000	Samsung 840
SLC (floating-gate) NOR	100,000 to 1,000,000	Numonyx M58BW (Endurance rating of 100,000 erases per block); Spansion S29CD016J (Endurance rating of 1,000,000 erases per block)
MLC (floating-gate) NOR	100,000	Numonyx J3 flash

However, by applying certain algorithms and design paradigms such as wear leveling and memory over-provisioning, the endurance of a storage system can be tuned to serve specific requirements.

In order to compute the longevity of the NAND flash, one must account for the size of the memory chip, the type of memory (e.g. SLC/MLC/TLC), and use pattern.

Flash file Systems

Because of the particular characteristics of flash memory, it is best used with either a controller to perform wear leveling and error correction or specifically designed flash file systems, which spread writes over the media and deal with the long erase times of NOR flash blocks. The basic concept behind flash file systems is the following: when the flash store is to be updated, the file system will write a new copy of the changed data to a fresh block, remap the file pointers, then erase the old block later when it has time.

In practice, flash file systems are used only for memory technology devices (MTDs), which are embedded flash memories that do not have a controller. Removable flash memory cards and USB flash drives have built-in controllers to perform wear leveling and error correction so use of a specific flash file system does not add any benefit.

Capacity

Multiple chips are often arrayed to achieve higher capacities for use in consumer electronic devices such as multimedia players or GPSs. The capacity of flash chips generally follows Moore"s Law because they are manufactured with many of the same integrated circuits techniques and equipment.

Consumer flash storage devices typically are advertised with usable sizes expressed as a small integer power of two (2, 4, 8, etc.) and a designation of megabytes (MB) or gigabytes (GB); e.g., 512 MB, 8 GB. This includes SSDs marketed as hard drive replacements, in accordance with traditional hard drives, which use decimal prefixes. Thus, an SSD marked as "64 GB" is at least $64 \times 1{,}000^3$ bytes (64 GB). Most users will have slightly less capacity than this available for their files, due to the space taken by file system metadata.

The flash memory chips inside them are sized in strict binary multiples, but the actual total capacity of the chips is not usable at the drive interface. It is considerably larger than the advertised capacity in order to allow for distribution of writes (wear leveling), for sparing, for error correction codes, and for other metadata needed by the device's internal firmware.

In 2005, Toshiba and SanDisk developed a NAND flash chip capable of storing 1 GB of data using multi-level cell (MLC) technology, capable of storing two bits of data per cell. In September 2005, Samsung Electronics announced that it had developed the world's first 2 GB chip.

In March 2006, Samsung announced flash hard drives with a capacity of 4 GB, essentially the same order of magnitude as smaller laptop hard drives, and in September 2006, Samsung announced an 8 GB chip produced using a 40 nm manufacturing process. In January 2008, SanDisk announced availability of their 16 GB MicroSDHC and 32 GB SDHC Plus cards.

More recent flash drives (as of 2012) have much greater capacities, holding 64, 128, and 256 GB.

A joint development at Intel and Micron will allow the production of 32 layer 3.5 terabyte (TB) NAND flash sticks and 10 TB standard-sized SSDs. The device includes 5 packages of 16 x 48 GB TLC dies, using a floating gate cell design.

Flash chips continue to be manufactured with capacities under or around 1 MB, e.g., for BIOS-ROMs and embedded applications.

In July 2016, Samsung announced the 4TB Samsung 850 EVO which utilizes their 256Gb 48-Layer TLC 3D V-NAND. In August 2016, Samsung announced a 32TB 2.5" SAS SSD based on their 512Gb 64-Layer TLC 3D V-NAND. Further, Samsung expects to unveil SSDs with up to 100TB of storage by 2020.

Transfer rates

Flash memory devices are typically much faster at reading than writing. Performance also depends on the quality of storage controllers which become more critical when devices are partially full. Even when the only change to manufacturing is die-shrink, the absence of an appropriate controller can result in degraded speeds.

Applications

Serial flash

Serial flash is a small, low-power flash memory that uses a serial interface, typically Serial Peripheral Interface Bus (SPI), for sequential data access. When incorporated into an embedded system, serial flash requires fewer wires on the PCB than parallel flash memories, since it transmits and receives data one bit at a time. This may permit a reduction in board space, power consumption, and total system cost.

There are several reasons why a serial device, with fewer external pins than a parallel device, can significantly reduce overall cost:

- Many ASICs are pad-limited, meaning that the size of the die is constrained by the number of wire bond pads, rather than the complexity and number of gates used for the device logic. Eliminating bond pads thus permits a more compact integrated circuit, on a smaller die; this increases the number of dies that may be fabricated on a wafer, and thus reduces the cost per die.
- Reducing the number of external pins also reduces assembly and packaging costs. A serial device may be packaged in a smaller and simpler package than a parallel device.
- Smaller and lower pin-count packages occupy less PCB area.
- Lower pin-count devices simplify PCB routing.

There are two major SPI flash types. The first type is characterized by small pages and one or more internal SRAM page buffers allowing a complete page to be read to the buffer, partially modified, and then written back (for example, the Atmel AT45 *DataFlash* or the Micron Technology Page Erase NOR Flash). The second type has larger sectors. The smallest sectors typically found in an SPI flash are 4 kB, but they can be as large as 64 kB. Since the SPI flash lacks an internal SRAM buffer, the complete page must be read out and modified before being written back, making it slow to manage. *SPI flash* is cheaper than *DataFlash* and is therefore a good choice when the application is code shadowing.

The two types are not easily exchangeable, since they do not have the same pinout, and the command sets are incompatible.

Firmware storage

With the increasing speed of modern CPUs, parallel flash devices are often much slower than the memory bus of the computer they are connected to. Conversely, modern SRAM offers access times below 10 ns, while DDR2 SDRAM offers access times below 20 ns. Because of this, it is often desirable to shadow code stored in flash into RAM; that is, the code is copied from flash into RAM before execution, so that the CPU may access it at full speed. Device firmware may be stored in a serial flash device, and then copied into SDRAM or SRAM when the device is powered-up. Using an external serial flash device rather than on-chip flash removes the need for significant process compromise (a process that is good for high-speed logic is generally not good for flash and vice versa). Once it is decided to read the firmware in as one big block it is common to add compression to allow a smaller flash chip to be used. Typical applications for serial flash include storing firmware for hard drives, Ethernet controllers, DSL modems, wireless network devices, etc.

Flash memory as a replacement for hard drives

Main article: Solid-state drive

One more recent application for flash memory is as a replacement for hard disks. Flash memory does not have the mechanical limitations and latencies of hard drives, so a solid-state drive (SSD) is attractive when considering speed, noise, power consumption, and reliability. Flash drives are gaining traction as mobile device secondary storage devices; they are also used as substitutes for hard drives in high-performance desktop computers and some servers with RAID and SAN architectures.

There remain some aspects of flash-based SSDs that make them unattractive. The cost per gigabyte of flash memory remains significantly higher than that of hard disks. Also flash memory has a finite number of P/E cycles, but this seems to be currently under control since warranties on flash-based SSDs are approaching those of current hard drives. In addition, deleted files on SSDs can remain for an indefinite period of time before being overwritten by fresh data; erasure or shred techniques or software that work well on magnetic hard disk drives have no effect on SSDs, compromising security and forensic examination.

For relational databases or other systems that require ACID transactions, even a modest amount of flash storage can offer vast speedups over arrays of disk drives.

In June 2006, Samsung Electronics released the first flash-memory based PCs, the Q1-SSD and Q30-SSD, both of which used 32 GB SSDs, and were at least initially available only in South Korea.

A solid-state drive was offered as an option with the first Macbook Air introduced in 2008, and from 2010 onwards, all Macbook Air laptops shipped with an SSD. Starting in late 2011, as part of Intel"s Ultrabook initiative, an increasing number of ultra thin laptops are being shipped with SSDs standard.

There are also hybrid techniques such as hybrid drive and ReadyBoost that attempt to combine the advantages of both technologies, using flash as a high-speed non-volatile cache for files on the disk that are often referenced, but rarely modified, such as application and operating system executable files.

Flash memory as RAM

As of 2012, there are attempts to use flash memory as the main computer memory, DRAM.

Archival or long-term storage

It is unclear how long flash memory will persist under archival conditions—i.e., benign temperature and humidity with infrequent access with or without prophylactic rewrite. Anecdotal evidence suggests that the technology is reasonably robust on the scale of years. Datasheets of Atmel's flash-based "ATmega" microcontrollers typically promise retention times of 20 years at 85 °C and 100 years at 25 °C.

An article from CMU in 2015 writes that "Today's flash devices, which do not require flash refresh, have a typical retention age of 1 year at room temperature." And that temperature can lower the retention time exponentially. The phenomenon can be modeled by Arrhenius law.

Industry

One source states that, in 2008, the flash memory industry includes about US$9.1 billion in production and sales. Other sources put the flash memory market at a size of more than US$20 billion in 2006, accounting for more than eight percent of the overall semiconductor market and more than 34 percent of the total semiconductor memory market. In 2012, the market was estimated at $26.8 billion.

Flash Scalability

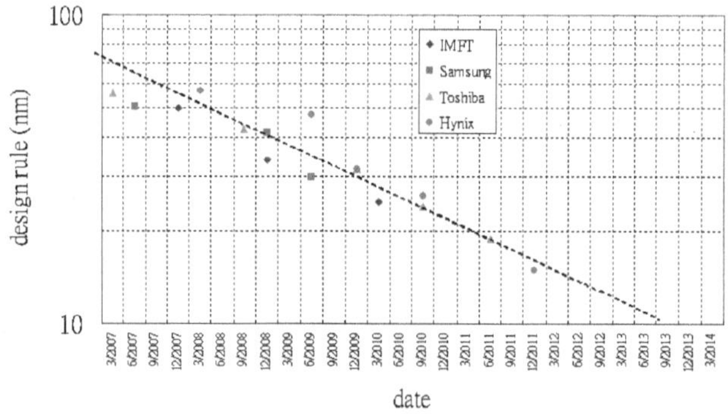

The aggressive trend of the shrinking process design rule or technology node in NAND flash memory technology effectively accelerates Moore's Law.

Due to its relatively simple structure and high demand for higher capacity, NAND flash memory is the most aggressively scaled technology among electronic devices. The heavy competition among the top few manufacturers only adds to the aggressiveness in shrinking the design rule or process technology node. While the expected shrink timeline is a factor of two every three years per original version of Moore's law, this has recently been accelerated in the case of NAND flash to a factor of two every two years.

ITRS or company	2010	2011	2012	2013	2014	2015	2016
ITRS Flash Roadmap 2011	32 nm	22 nm	20 nm	18 nm	16 nm		
Updated ITRS Flash Roadmap					17 nm	15 nm	14 nm

Samsung Samsung 3D NAND	35-32 nm	27 nm	21 nm (MLC, TLC)	19 nm	19-16 nm V-NAND (24L)	12 nm V-NAND (32L)	12 nm
Micron, Intel	34-25 nm	25 nm	20 nm (MLC + HKMG)	20 nm (TLC)	16 nm	3D-NAND	3D-NAND Gen2
Toshiba, Sandisk	43-32 nm	24 nm	19 nm (MLC, TLC)		15 nm	3D NAND BiCS	3D NAND BiCS
SK Hynix	46-35 nm	26 nm	20 nm (MLC)		16 nm		

As the feature size of flash memory cells reaches the minimum limit, further flash density increases will be driven by greater levels of MLC, possibly 3-D stacking of transistors, and improvements to the manufacturing process. The decrease in endurance and increase in uncorrectable bit error rates that accompany feature size shrinking can be compensated by improved error correction mechanisms. Even with these advances, it may be impossible to economically scale flash to smaller and smaller dimensions as the number of electron holding capacity reduces. Many promising new technologies (such as FeRAM, MRAM, PMC, PCM, ReRAM, and others) are under investigation and development as possible more scalable replacements for flash.

REFERENCES

1. "A Flash Storage Technical and Economic Primer". flashstorage.com. 30 March 2015.
2. "A Survey of Software Techniques for Using Non-Volatile Memories for Storage and Main Memory Systems", S. Mittal and J. Vetter, IEEE TPDS, 2015
3. "Owners of QM2 seabed camera found". BBC News. 11 February 2010.
4. Fulford, Benjamin (24 June 2002). "Unsung hero". Forbes. Retrieved 18 March 2008.
5. US 4531203 Fujio Masuoka
6. "NAND Flash Memory: 25 Years of Invention, Development - Data Storage - News & Reviews - eWeek.com". eweek.com.
7. Masuoka, F.; Momodomi, M.; Iwata, Y.; Shirota, R. (1987). "New ultra high density EPROM and flash EEPROM with NAND structure cell". Electron Devices Meeting, 1987 International. IEEE. Retrieved 4 January 2013.
8. Tal, Arie (February 2002). "NAND vs. NOR flash technology: The designer should weigh the options when using flash memory". Retrieved 31 July 2010.
9. "H8S/2357 Group, H8S/2357F-ZTATTM, H8S/2398F-ZTATTM Hardware Manual, Section 19.6.1" (PDF). Renesas. October 2004. Retrieved 23 January 2012. The flash memory can be reprogrammed up to 100 times.
10. "AMD DL160 and DL320 Series Flash: New Densities, New Features" (PDF). AMD. July 2003. Retrieved 13 November 2014. The devices offer single-power-supply operation (2.7 V to 3.6 V), sector architecture, Embedded Algorithms, high performance, and a 1,000,000 program/erase cycle endurance guarantee.
11. "SanDisk Unveils the World"s Highest Capacity microSD™ Card (200GB)". Sandisk. 1 March 2015. Retrieved 1 April 2015.
12. Basinger, Matt (18 January 2007), PSoC Designer Device Selection Guide (PDF), AN2209, The PSoC ... utilizes a unique Flash process: SONOS

13. Yasufuku, Tadashi; Ishida, Koichi; Miyamoto, Shinji; Nakai, Hiroto; Takamiya, Makoto; Sakurai, Takayasu; Takeuchi, Ken (2009), Inductor design of 20-V boost converter for low power 3D solid state drive with NAND flash memories, pp. 87–92(abstract).
14. Micheloni, Rino; Marelli, Alessia; Eshghi, Kam (2012), Inside Solid State Drives (SSDs)
15. Micheloni, Rino; Crippa, Luca (2010), Inside NAND Flash Memories In particular, pp 515-536: K. Takeuchi. "Low power 3D-integrated SSD"
16. Mozel, Tracey (2009), CMOSET Fall 2009 Circuits and Memories Track Presentation Slides
17. Tadashi Yasufuku et al., "Inductor and TSV Design of 20-V Boost Converter for Low Power 3D Solid State Drive with NAND Flash Memories". 2010.
18. Hatanaka, T. and Takeuchi, K. "4-times faster rising VPASS (10V), 15% lower power VPGM (20V), wide output voltage range voltage generator system for 4-times faster 3D-integrated solid-state drives". 2011.
19. Takeuchi, K., "Low power 3D-integrated Solid-State Drive (SSD) with adaptive voltage generator". 2010.
20. Ishida, K. et al., "1.8 V Low-Transient-Energy Adaptive Program-Voltage Generator Based on Boost Converter for 3D-Integrated NAND Flash SSD". 2011.
21. A. H. Johnston, "Space Radiation Effects in Advanced Flash Memories". NASA Electronic Parts and Packaging Program (NEPP). 2001. "... internal transistors used for the charge pump and erase/write control have much thicker oxides because of the requirement for high voltage. This causes flash devices to be considerably more sensitive to total dose damage compared to other ULSI technologies. It also implies that write and erase functions will be the first parameters to fail from total dose. ... Flash memories will work at much higher radiation levels in the read mode. ... The charge pumps that are required to generate the high voltage for erasing and writing are usually the most sensitive circuit functions, usually failing below 10 krad(SI)."
22. Zitlaw, Cliff. "The Future of NOR Flash Memory". Memory Designline. UBM Media. Retrieved 3 May 2011.
23. "Samsung moves into mass production of 3D flash memory". Gizmag.com. Retrieved 2013-08-27.
24. Jonathan Thatcher, Fusion-io; Tom Coughlin, Coughlin Associates; Jim Handy, Objective-Analysis; Neal Ekker, Texas Memory Systems (April 2009). "NAND Flash Solid State Storage for the Enterprise, An In-depth Look at Reliability" (PDF). Solid State Storage Initiative (SSSI) of the Storage Network Industry Association (SNIA). Retrieved 6 December 2011.
25. «Micron Collaborates with Sun Microsystems to Extend Lifespan of Flash-Based Storage, Achieves One Million Write Cycles" (Press release). Micron Technology, Inc. 17 December 2008.
26. "Taiwan engineers defeat limits of flash memory". phys.org.
27. "Flash memory made immortal by fiery heat". theregister.co.uk.
28. "Flash memory breakthrough could lead to even more reliable data storage."

29. "TN-29-17 NAND Flash Design and Use Considerations Introduction" (PDF). Micron. April 2010. Retrieved 29 July 2011.
30. Kawamatus, Tatsuya. "TECHNOLOGY FOR MANAGING NAND FLASH" (PDF). Hagiwara sys-com co., LTD. Retrieved 1 August 2011.
31. Richard Blish. "Dose Minimization During X-ray Inspection of Surface-Mounted Flash ICs". p. 1.
32. Richard Blish. "Impact of X-Ray Inspection on Spansion Flash Memory".
33. "SanDisk Extreme PRO SDHC/SDXC UHS-I Memory Card". Retrieved 2016-02-03.
34. "Samsung 32GB USB 3.0 Flash Drive FIT MUF-32BB/AM". Retrieved 2016-02-03.
35. Spansion. "What Types of ECC Should Be Used on Flash Memory?". 2011.
36. "DSstar: TOSHIBA ANNOUNCES 0.13 MICRON 1GB MONOLITHIC NAND". Tgc.com. 2002-04-23. Retrieved 2013-08-27.
37. Kim, Jesung; Kim, John Min; Noh, Sam H.; Min, Sang Lyul; Cho, Yookun (May 2002). "A Space-Efficient Flash Translation Layer for CompactFlash Systems" (PDF). Proceedings of the IEEE. **48** *(2). pp. 366–375. Retrieved 2008-08-15.*
38. TN-29-07: Small-Block vs. Large-Block NAND flash Devices Explains 512+16 and 2048+64-byte blocks
39. AN10860 LPC313x NAND flash data and bad block management Explains 4096+128-byte blocks.
40. Thatcher, Jonathan (18 August 2009). "NAND Flash Solid State Storage Performance and Capability – an In-depth Look" (PDF). SNIA. Retrieved 2012-08-28.
41. "Samsung ECC algorithm" (PDF). Samsung. June 2008. Retrieved 15 August 2008.
42. "Open NAND Flash Interface Specification" (PDF). Open NAND Flash Interface. 28 December 2006. Retrieved 31 July 2010.
43. A list of ONFi members is available at http://onfi.org/membership/.
44. "Toshiba Introduces Double Data Rate Toggle Mode NAND In MLC And SLC Configurations". toshiba.com.
45. "Dell, Intel And Microsoft Join Forces To Increase Adoption Of NAND-Based Flash Memory In PC Platforms". REDMOND, Wash: Microsoft. 30 May 2007. Retrieved 12 August 2014.
46. See pages 5–7 of Toshiba's "NAND Applications Design Guide" under External links.
47. NAND Flash 101: An Introduction to NAND Flash and How to Design It In to Your Next Product (PDF), Micron, pp. 2–3, TN-29-19
48. Pavan, Paolo; Bez, Roberto; Olivo, Piero; Zanoni, Enrico (1997). "Flash Memory Cells – An Overview" (PDF). Proceedings of the IEEE. **85** *(8) (published August 1997). pp. 1248–1271. doi:10.1109/5.622505. Retrieved 15 August 2008.*
49. "NAND Evolution and its Effects on Solid State Drive Useable Life" (PDF). Western Digital. 2009. Retrieved 22 April 2012.
50. "A survey of address translation technologies for flash memories", ACM Computing Surveys, 2014.

51. "Flash vs DRAM follow-up: chip stacking". The Daily Circuit. 22 April 2012. Retrieved 22 April 2012.
52. http://www.convertunits.com/type/computer+data+storage
53. Shilov, Anton (12 September 2005). "Samsung Unveils 2GB Flash Memory Chip". X-bit labs. Retrieved 30 November 2008.
54. Gruener, Wolfgang (11 September 2006). "Samsung announces 40 nm Flash, predicts 20 nm devices". TG Daily. Retrieved 30 November 2008.
55. "SanDisk Media Center". sandisk.com.
56. http://www.pcworld.com/businesscenter/article/225370/look_out_for_the_256gb_thumb_drive_and_the_128gb_tablet.html; http://techcrunch.com/2009/07/20/kingston-outs-the-first-256gb-flash-drive/ 20 July 2009, Kingston DataTraveler 300 is 256 GB.
57. Borghino, Dario (March 31, 2015). "3D flash technology moves forward with 10 TB SSDs and the first 48-layer memory cells". Gizmag. Retrieved March 2015. Check date values in: |access-date= (help)
58. "Samsung Launches Monster 4TB 850 EVO SSD Priced at $1,499 | Custom PC Review". Custom PC Review. 2016-07-13. Retrieved 2016-10-08.
59. "Samsung Unveils 32TB SSD Leveraging 4th Gen 64-Layer 3D V-NAND | Custom PC Review". Custom PC Review. 2016-08-11. Retrieved 2016-10-08.
60. Master, Neal; Andrews, Mathew; Hick, Jason; Canon, Shane; Wright, Nicholas (2010). "Performance analysis of commodity and enterprise class flash devices" (PDF). IEEE Petascale Data Storage Workshop.
61. "DailyTech - Samsung Confirms 32nm Flash Problems, Working on New SSD Controller". dailytech.com.
62. Many serial flash devices implement a bulk read mode and incorporate an internal address counter, so that it is trivial to configure them to transfer their entire contents to RAM on power-up. When clocked at 50 MHz, for example, a serial flash could transfer a 64 Mbit firmware image in less than two seconds.
63. Lyth0s (17 March 2011). "SSD vs. HDD". elitepcbuilding.com. Retrieved 11 July 2011.
64. "Flash Solid State Disks – Inferior Technology or Closet Superstar?". STORAGEsearch. Retrieved 30 November 2008.
65. Vadim Tkachenko. "Intel SSD 910 vs HDD RAID in tpcc-mysql benchmark". MySQL Performance Blog.
66. Matsunobu, Yoshinori. "SSD Deployment Strategies for MySQL." Sun Microsystems, 15 April 2010.
67. "Samsung Electronics Launches the World"s First PCs with NAND Flash-based Solid State Disk". Press Release. Samsung. 24 May 2006. Retrieved 30 November 2008.
68. Douglas Perry (2012) Princeton: Replacing RAM with Flash Can Save Massive Power.
69. "8-Bit AVR Microcontroller ATmega32A Datasheet Complete" (PDF). 2016-02-19. p. 18. Retrieved 2016-05-29. Reliability Qualification results show that the projected data retention failure rate is much less than 1 PPM over 20 years at 85°C or 100 years at 25°C
70. "Data Retention in MLC NAND Flash Memory: Characterization, Optimization, and Recovery" (PDF). 2015-01-27. p. 10. Retrieved 2016-04-27.

71. Yinug, Christopher Falan (July 2007). "The Rise of the Flash Memory Market: Its Impact on Firm Behavior and Global Semiconductor Trade Patterns" (PDF). Journal of International Commerce and Economics. Archived from the original (PDF) on 29 May 2008. Retrieved 19 April 2008.
72. NAND memory market rockets, April 17, 2013, Nermin Hajdarbegovic, TG Daily, retrieved at 18 April 2013
73. "Technology Roadmap for NAND Flash Memory". techinsights. April 2013. Archived from the original on 9 January 2015. Retrieved 9 January 2015.
74. "Technology Roadmap for NAND Flash Memory". techinsights. April 2014. Archived from the original on 9 January 2015. Retrieved 9 January 2015.
75. Lal Shimpi, Anand (2 December 2010). "Micron"s ClearNAND: 25nm + ECC, Combats Increasing Error Rates". Anandtech. Retrieved 2 December 2010.
76. Kim, Kinam; Koh, Gwan-Hyeob (16 May 2004). Future Memory Technology including Emerging New Memories(PDF). Serbia and Montenegro: Proceedings of the 24th International Conference on Microelectronics. pp. 377–384. Retrieved 2008-08-15.
77. FPGA Architecture for the Challenge". toronto.edu.
78. Wisniewski, Remigiusz (2009). Synthesis of compositional microprogram control units for programmable devices. Zielona Góra: University of Zielona Góra. p. 153. ISBN 978-83-7481-293-1.
79. «FPGA Signal Integrity tutorial". altium.com.
80. NASA: FPGA drive strength
81. Mike Thompson. "Mixed-signal FPGAs provide GREEN POWER". EE Times, 2007-07-02.
82. History of FPGAs at the Wayback Machine (archived April 12, 2007)
83. Google Patent Search, "Re-programmable PLA". Retrieved February 5, 2009.
84. Google Patent Search, "Dynamic data re-programmable PLA". Retrieved February 5, 2009.
85. "In the Beginning". altera.com. 21 April 2015.
86. "XCELL issue 32" (PDF). Xilinx.
87. Peter Clarke, EE Times, "Xilinx, ASIC Vendors Talk Licensing." June 22, 2001. Retrieved February 10, 2009.
88. Funding Universe. "Xilinx, Inc." Retrieved January 15, 2009.
89. Clive Maxfield, Programmable Logic DesignLine, "Xilinx unveil revolutionary 65nm FPGA architecture: the Virtex-5 family. May 15, 2006. Retrieved February 5, 2009.
90. Press Release, "Xilinx Co-Founder Ross Freeman Honored as 2009 National Inventors Hall of Fame Inductee for Invention of FPGA"
91. US 4870302, Freeman, Ross H., "Configurable electrical circuit having configurable logic elements and configurable interconnects", published 19 February 1988, issued 26 September 1989
92. "Top FPGA Companies For 2013". sourcetech411.com.
93. Maxfield, Clive (2004). The Design Warrior"s Guide to FPGAs: Devices, Tools and Flows. Elsevier. p. 4. ISBN 978-0-7506-7604-5.

94. McConnel, Toni. EE Times. "ESC – Xilinx All Programmable System on a Chip combines best of serial and parallel processing." April 28, 2010. Retrieved February 14, 2011.
95. Nass, Rich, EE Times. "Xilinx puts ARM core into its FPGAs." April 27, 2010. Retrieved February 14, 2011.
96. Wilson, Richard, Electronics Weekly. "Xilinx acquires ESL firm to make FPGAs easier to use." January 31, 2011. Retrieved February 15, 2011.
97. "Microsoft Supercharges Bing Search With Programmable Chips". WIRED. 16 June 2014.
98. Dylan McGrath, EE Times, "FPGA Market to Pass $2.7 Billion by ‹10, In-Stat Says". May 24, 2006. Retrieved February 5, 2009.
99. "Global FPGA Market Analysis And Segment Forecasts To 2020 – FPGA Industry, Outlook, Size, Application, Product, Share, Growth Prospects, Key Opportunities, Dynamics, Trends, Analysis, FPGA Report – Grand View Research Inc". grandviewresearch.com.
100. Dylan McGrath, EE Times, "Gartner Dataquest Analyst Gives ASIC, FPGA Markets Clean Bill of Health". June 13, 2005. Retrieved February 5, 2009.
101. Virtex-4 Family Overview
102. Kuon, Ian; Rose, Jonathan (2006). "Measuring the gap between FPGAs and ASICs". Proceedings of the international symposium on Field programmable gate arrays – FPGA"06. New York, NY: ACM. pp. 21–30. doi:10.1145/1117201.1117205. ISBN 1-59593-292-5.
103. Tim Erjavec, White Paper, "Introducing the Xilinx Targeted Design Platform: Fulfilling the Programmable Imperative." February 2, 2009. Retrieved February 2, 2009
104. Huffmire Paper "Managing Security in FPGA-Based Embedded Systems." Nov-Dec 2008. Retrieved Sept 22, 2009
105. "EETimes on PUF: Security features for non-security experts – Intrinsic ID". Intrinsic ID.
106. "Publications and Presentations". bdti.com.
107. "Xilinx aims 65-nm FPGAs at DSP applications". EETimes.
108. "CrypTech: Building Transparency into Cryptography t"(PDF).
109. "CrypTech: Building Transparency into Cryptography t"(PDF).
110. http://www.altera.com/literature/hb/cyc2/cyc2_cii51002.pdf
111. "Documentation: Stratix IV Devices" (PDF). Altera.com. 2008-06-11. Retrieved 2013-05-01.
112. http://www.xilinx.com/support/documentation/user_guides/ug070.pdf
113. Dean Takahashi, VentureBeat. "Intel connection helped chip startup Tabula raise $108M." May 2, 2011. Retrieved May 13, 2011.
114. Lawrence Latif, The Inquirer. "FPGA manufacturer claims to beat Moore"s Law." October 27, 2010. Retrieved May 12, 2011.
115. EDN Europe. "Xilinx adopts stacked-die 3D packaging." November 1, 2010. Retrieved May 12, 2011.
116. http://www.xilinx.com/support/documentation/white_papers/wp380_Stacked_Silicon_Interconnect_Technology.pdf

117. “Intel Custom Foundry EMIB”. Intel.
118. “Why use OpenCL on FPGAs?”. StreamComputing.
119. Edwards, John (June 1, 2006). “No room for Second Place: Xilinx and Altera slug it out for supremacy in the changing PLD market”. EDN. Retrieved May 11, 2012.
120. “Altera and Xilinx Report: The Battle Continues”. Seeking Alpha. July 17, 2008. Retrieved November 13, 2013.
121. “Xilinx ISE WebPACK”.
122. “Quartus II Web edition software”.
123. “Achronix to use Intel”s 22 nm manufacturing”.
124. “Top FPGA Companies For 2013”.
125. “Tabula”s Time Machine — Micro Processor Report”(PDF).
126. Tabula to shut down; 120 jobs lost at fabless chip company Silicon Valley Business Journal
127. “Intel to buy Altera for $16.7 billion in its biggest deal ever”. Reuters.

Index

A

Actual calculation 187
Analog-to-digital converters (ADCs) 201
Application-specific integrated circuit (ASIC) 199
Appropriate small 179
Automatic test pattern generation (ATG) 153

B

Body mass index (BMI) 172

C

Configurable logic blocks (CLBs) 202

D

Digital-to-analog converters (DACs) 201
Discrete Incremental Clustering (DIC) 52
Distribution function 186

E

Error analysis 185, 187
Extensions for Scientific Computation (XSC) 188

F

Field-programmable gate array (FPGA) 199
Field Programmable Gate Arrays (FPGA) 153
Field-programmable gate arrays (FPGAs) 201
Fluctuation range 173, 186
Footprint of uncertainty (FOU) 47
Fuzzy Kohonen Partitioning (FKP) 52
Fuzzy Neural Network (FNN) 51

I

Interval algebra 187, 189
Interval numbers 178

L

Learning Vector Quantization (LVQ) 52
Lower membership function (LMF) 47

M

Meaningful conclusions 181

N

Neuro-Fuzzy System (NFS) 51

O

Ordered weighted averaging (OWA) 192

P

Programmable Logic Arrays (PLAs) 153
Programmable logic devices (PLDs) 201
Programmable read-only memory (PROM) 201

R

Rational matrix 182

S

Solid-state drive (SSD) 242

T

Takagi-Sugeno-Kang (TSK) 51